Winning at New Products

W$INNING AT NEW PRODUCTS

Robert G. Cooper

Addison-Wesley Publishing Company, Inc.
Reading, Massachusetts • Menlo Park, California • New York
Don Mills, Ontario • Wokingham, England • Amsterdam
Bonn • Paris • Milan • Madrid • Sydney • Singapore
Tokyo • Seoul • Taipei • Mexico City • San Juan

The publisher offers discounts on this book when ordered in quantity for special sales. For more information please contact:
Corporate & Professional Publishing Group
Addison-Wesley Publishing Company
Route 128
Reading, Massachusetts 01867

Library of Congress Cataloging-in-Publication Data

Cooper, Robert G.
Winning at new products/ Robert G. Cooper
p.   cm
Bibliography: p.
Includes index.
ISBN 0-201-12038-0
1. New products--Management. 2. New products--Marketing
3. Creative ability in business. 4. New business enterprises.
I. Title   HD69.N4C658   1988   658.5'75--dc 19   88-921

Cover design by Mike Fender

Text design by Brant Cowie/Artplus

Typesetting by Lithocomp Limited

Fifth printing,September 1991
First paperback printing, April 1988
5 6 7 8 9 10 DO 9594939291
ISBN 0-201-12038-0

To Linda, Barbara, and Heather.

Contents

PREFACE   *ix*

CHAPTER 1   New Products: The Game   *1*

CHAPTER 2   What Separates the Winners from the Losers?   *15*

CHAPTER 3   The New-Product Game Plan   *35*

CHAPTER 4   Finding Good New-Product Ideas   *67*

CHAPTER 5   Picking the Right New-Product Project   *95*

CHAPTER 6   Defining the Product Concept and Specifying the Product   *123*

CHAPTER 7   Testing the Product and the Strategy   *161*

CHAPTER 8   The Final Play: Into the Market   *183*

CHAPTER 9   The Long Term: What Markets, Products, and Technologies?   *217*

APPENDIX A   The NewProd Screening Model   *247*

APPENDIX B   Benefit-Contribution Screening Methods: Financial Indices   *253*

APPENDIX C   Survey Methods   *257*

NOTES   *262*

INDEX   *270*

Preface

This book is about product innovation — the conception, development, and commercialization of a new product. The topic has generated widespread interest in recent years. Technological capability, high-technology growth industries, R & D spending, and the threat of emergent technologies from other countries are discussed in countless company strategy meetings.

Only in recent years, however, has the manager been able to find much in the way of outside help in managing new products. Until recently, our knowledge about how products are developed and launched was very limited. There was little reading material on new-product development available to the manager. Most of the books on the topic written in the last decade have been aimed at university-level courses on the management of technology and new products. Because they are university texts, they were often inappropriate for managers — too long, too theoretical, too descriptive, and lacking in an action orientation.

This book is written for the practicing manager. The manager with years of experience in new products will recognize many familiar concepts and tools integrated and expressed, perhaps in new ways. For him or her, the book is a "pulling together" of these concepts. For the manager new to product innovation, the book is an excellent how-to guide — a step-by-step blueprint to conceiving, developing, and launching new products.

The book has been written both for technical people and for marketing and business people. For the person with an engineering or physical science background, the book portrays a side of the innovation process that is sometimes misunderstood, and often dismissed, by technical people. For the marketer, product manager, or business manager, the book attempts to provide a framework and introduce discipline into what is often an ad hoc area of business.

A number of people helped me in the writing of this book. Many of my present and former colleagues discussed issues and ideas that later found their way into the book. Professors Roger Bennett, Blair Little, Roger More, Roger Calantone, Ken Deal, and Elko Kleinschmidt helped me clarify many of the thoughts in the book. My business contacts, in particular Bob Davis, Dan Ennis, Peter Josty, Jim Koziak, Ron Body, and Gerry Buelow, provided many suggestions and examples. I also owe many thanks to those business people who took many hours to review the original manuscript and helped to fashion the final product: David Sorensen, Nirmal Pujari, Dan Ennis, Rob Bracey, Anne Carlyle, and Don Morrison. Finally, special thanks go to two people who spent an enormous amount of time with me on the project and on the manuscript itself — Jocelyn Klemm, Business Editor, Professional Division of Holt, Rinehart & Winston of Canada Ltd., and Kathy Johnson, who did the copy editing.

New Products:
The Game

The Players, Teams, and Arenas

The development of new products is a high-risk and potentially very rewarding game. The teams are the many corporations who commit manpower and money to new products in the hope of becoming winners. Some of the more successful teams are well known: IBM, Xerox, 3M, DuPont, GE, Procter & Gamble, and others who, year after year, have a steady stream of new-product winners. There are also new teams in the league; a few of these, such as Apple Computers, seem to come out of nowhere and strike it rich with an extremely successful new product.

The game has its individual players and stars: researchers and scientists, engineers and designers, sales and marketing people, product managers, and even production, finance, and accounting types. The coaching staff are the managers. Sometimes, with luck, the players really do play as a team. All too often, though, team cohesion and team discipline are missing — and the game is reduced to a one-person or one-department show; the integrated effort needed to win the game is lacking. At times it seems that half the team doesn't even realize that there's a game going on.

The game also has its playing fields or arenas: markets, technology areas, and business areas where the game is played. Some teams appear to pick arenas in which they have a better chance of winning, and in which they can build on their distinctive strengths. Other teams are consistently in the wrong place at the wrong time. Still others aren't even aware that they have a choice of arenas, and drift into different fields almost by chance. The selection of the arena is the strategic facet of the new-product game.

Finally, as in any game, there are winners and losers. The winners are the companies whose sales and profits have grown steadily over the decades, outperforming the rest of the industry, the ones who boast an enviable portfolio of successful new products that have a major impact on the corporation's fortunes. Even the most successful teams are bound to lose a few games, however, and every firm has its list of new-product losers. But too many companies have more than their fair share of failures. The results are often disastrous; some of these teams disappear into the minor leagues, and others fail to make it to the next season.

The Game Plan

This book is about mastering the new-product game. It offers a game plan designed to increase the odds of winning at new products. The game plan is based on the experiences, approaches, and methods of a number of successful corporations — companies who have a winning record in the new-product game — and of some not-so-successful ones. Some of the examples are taken from my own consulting practice (in these cases I have altered names and identifying details). Other illustrations are based on my own original research and on results that have been published elsewhere.

A word of caution: no game plan is foolproof. No single plan, no matter how carefully devised, works for all teams all the time. Although there will be strong similarities in the ideal game plans for different companies, there will also be differences — each firm's game plan will require some individual tailoring. And even the best plans are worthless if the execution is bad, if the players lack dedication and discipline, or if, as the sportscasters say, the team "lacks concentration."

This book has deliberately been kept short, simple, and on target. This is not a theoretical treatise on innovation management, nor is it an academic text or an encyclopedia. Rather, the book focuses on the most basic elements of winning. As is so often the case, many of these elements are childishly simple. Some suggestions may seem self-evident; but before becoming too complacent, ask yourself one question: if the answers are obvious, why is it that so few companies have seen the light?

Consider a classic case of a product failure.[1] This example is a classic in the sense that just about everything relating to the project was handled badly. The sad part of the tale is that this product should have been a winner, and a big one. As you identify the errors that were made, start developing your own checklist of the components of a winning game plan.

A Formula for Failure

"Our feelings ranged from a belief that customers would break down our door to take the device from our hands to the opinion of the engineers that the product was almost too good to be put on the market." This comment from the national sales manager of Gemini Inc. (not the company's real name) typified management's confidence in the new product. In 1969 Gemini, a large divisionalized manufacturer of electronic and electrical products, developed the Gemini 2000 — a high-speed electrostatic computer printer that operated four times faster than the fastest IBM line printer then on the market, and sold for one-quarter of the price.

The idea for the 2000 was born almost by accident. Basic research led to the conclusion that electrostatic technology could be used to create a printed page. The result would be a very high-speed printer with few moving parts. Development work on the electrostatic printer began in 1963. This was Gemini's first attempt at developing and marketing a computer-peripheral product. Work began slowly, with only two engineers on the development team. By 1966, two more engineers had been added, but no formal budget had been established. Management's attitude was, "Just work on it and see what can be done."

The Market

By 1969 the computer-peripheral market was large and rapidly growing. The market was also a highly competitive one, featuring a number of major manufacturers, such as IBM, Honeywell, RCA, CDC, and NCR, as well as numerous small specialty manufacturers. Even though Gemini was new to the market, little market research was undertaken prior to or during the development of the new product. "The engineers relied more on their own feelings," one executive observed. "Maybe they talked to some of the people who service our own computer; but on the whole, contacts with the computer industry were minimal."

By 1969, the product was ready for introduction. But still little market information had been gathered, and no formal marketing plan for the product had been developed. The national sales manager explained that his was a small division that lacked the resources to do much marketing planning and research. In spite of the lack of concrete evidence, however, management remained convinced that there was a market for its high-speed printer.

The Product

In 1969 the printer was introduced. Management forecast sales from five major market areas — computer line printers, CRT hard-copy printers, CRT microfilm hard-copy printers, teletype printers, and computer plotters. The initial market response was disappointing.

The 2000 had a printing speed of about one page per second, and was priced at about one-quarter of the price of the fastest line printers on the market. Additionally, it could print both alphanumeric characters and graphics. But there were many drawbacks to the final product. First, it required specially treated paper, which was forty times the cost of regular paper. A subsequent value-in-use calculation revealed that a heavy user of line printers would use about one million pages of paper per month at an extra $40,000 per month in paper costs — about three times the price of the printer. In addition, the printer could not make multiple copies, nor could it use preprinted forms. The paper size was wrong, too — the design team chose 8½-by-11-inch paper, when most users required larger, 14-inch wide paper. The paper feed accommodated rolled paper rather than fan-folded, and the paper roll was 300 feet long, which was good for only about five minutes of printer operation. Frequent roll changes meant that the machine was down a good part of the time.

To make a long and sad story short, the project limped along for many years. Millions of dollars were spent, but the printer never became a success.

What went wrong? The technology was brilliant, but the product was a dud. Although the printer was a technological breakthrough, it offered few real benefits to users in specific market segments. Remember that technology is the know-how that goes into the product; but a product must be viewed as "a bundle of benefits to the user." Seen in this light, the 2000 fared poorly. For users of high-speed computer printers, the prohibitive cost of specially treated paper, for example, far outweighed the economic advantages of the new product's high speed and low price. Second, a number of design deficiencies, such as the paper size and paper feed, were uncorrected even by the time the printer was introduced. Third, the target market was new to the company: Gemini didn't know much about the product's users and competitors; adequate market research was not done; and the launch effort itself was poorly planned and badly executed.

The Failure

Evidence shows that many of the problems were management's fault. With an effective game plan, the printer could have and should have been a great success, especially in the light of its timing. Some steps might have been taken to turn this disaster into a winner:

- an effective early screening and evaluation of the project to identify the main problem areas and to assist in the design a plan of action to minimize or overcome these potential difficulties;
- market studies, early in the development process, to identify the potential applications and market segments for the 2000 and to define user needs and preferences;
- an analysis of the strengths and weaknesses of competing products;
- the definition of the target market(s) and the specification of user-based design requirements prior to full-scale product development (perhaps several models aimed at different market segments would have resulted);
- the development of a sound marketing plan much earlier in the process;
- the undertaking of reviews and evaluations at critical stages in the new-product development process; and
- commitment and involvement on the part of management to ensure that the project was undertaken properly and with sufficient resources.

Many analyses and activities that are crucial to new-product success were omitted in the Gemini 2000 project. The company simply lacked an effective game plan. This lack became even more apparent when the company tackled a project outside its existing market and product boundaries.

What's "New" About a New Product?

There are many different types of "new" products. "Newness" can be defined in two senses:

- new to the company, in the sense that the firm has never made or sold this type of product before, although other firms might have;
- new to the market, or "innovative": the product is the first of its kind.

Viewed on a two-dimensional map, as shown in Exhibit 1.1, six different types or classes of new products can be identified:

Exhibit 1.1.   Categories of New Products

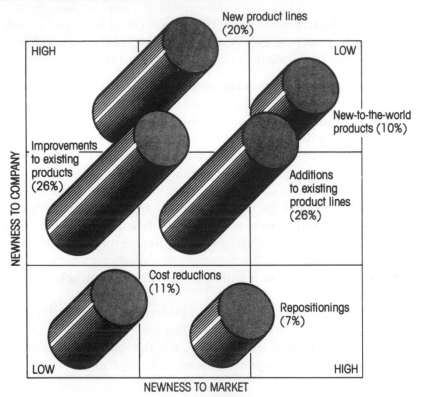

(Height of cylinder denotes number of introductions relative to total)

- **New-to-the-world products:** These new products are the first of their kind and create an entirely new market. This category represents only 10 per cent of all new products, according to a Booz Allen & Hamilton study.[2] Examples include the original Xerox copier, the first word processor, and the first home videocassette recorder.
- **New product lines:** These products, although not new to the marketplace, nonetheless are foreign to a particular firm. They allow a company to enter an established market for the first time. Procter & Gamble's entry into the cookie market is a good example: Duncan Hines' chocolate-chip cookies ("with the crispy exterior and chewy interior"), while a significant improvement over previous chocolate-chip cookies, were not a true innovation; but the product

category was new to Procter & Gamble. About 20 per cent of new products fit into this category.

- **Additions to existing product lines:** These are new to the firm, although not necessarily new to the marketplace. The difference between these products and new product lines is that these fit into a firm's existing product lines. They represent one of the largest categories of new products — about 26 per cent of all new-product launches.
- **Improvements and revisions to existing products:** These "not-so-new" products are essentially replacements of existing products in a firm's product line. They offer improved performance or greater perceived value over the "old" product. These "new and improved" products make up 26 per cent of new-product launches.
- **Repositionings:** These are essentially new applications for existing products, and often involve the retargeting of an old product to a new market segment. Johnson & Johnson's repositioning of their baby shampoo to the adult market as an adult shampoo and Arm & Hammer's repositioning of baking soda as a household deodorizer are examples. Repositionings account for about 7 per cent of all new products.
- **Cost reductions:** These are the least "new" of all product categories; they are new products designed to replace existing products in the firm's line; they yield similar performance and benefits at a lower cost. They represent 11 per cent of all new-product launches.

Most firms feature a mixed portfolio of new products. The two most popular categories, additions to the line and product improvements or revisions, are common to almost all firms, according to Booz Allen & Hamilton. The "step-out" products — new-to-the-world products, and new-to-the-firm product lines — constitute only 30 per cent of all new-product launches, but represent 60 per cent of the products viewed as "most successful."

Many firms steer clear of these two categories: 50 per cent of firms introduce no new-to-the-world products, and another 25 per cent develop no new product lines. This aversion to "step-out" and higher-risk products varies somewhat by industry, with the higher-technology industries relying much more on innovative products.

In this book, the term "new products" includes all of the categories listed above. But the fact is that most new products are not innovations; the typical new-product manager spends most of his or her time on the ordinary — the line modifications, extensions, and additions.

Suggestion: Conduct an audit or review of your firm's new-product portfolio, much as you would for an investment portfolio. Key questions to consider are:

- What proportion of your new products fits into each of the categories listed above?
- Is your portfolio sufficiently balanced? What are your firm's overall objectives for your new-product program — status quo, growth, or diversification? Given these objectives, what is the ideal new-product portfolio?
- What types of new products would help you to move closer to the ideal portfolio?

New Products: The Key to Corporate Prosperity

New-product development is one of the riskiest and most important activities of the modern corporation. It is easy to lose the game: most of us have witnessed experiences similar to that of the Gemini 2000. But before dismissing new products as being too chancy, consider the strategic importance of product innovation to the modern corporation.

- In 1982, an estimated $60 billion was spent on industrial R & D in the United States, or about 2.65 per cent of the GNP. This figure does not include many other new-product-related expenditures, such as business analysis, market studies, and commercialization costs. In Japan, R & D spending averages 2.47 per cent of the GNP; for West Germany, the figure is 2.81 per cent.
- For some industries, the figure is much higher: 18.3 per cent of sales in aircraft and missiles; 7.8 per cent in electrical equipment; and 8.9 per cent in scientific instruments. Exhibit 1.2 gives a breakdown of R & D expenditures as a percentage of company sales by industry. This heavy spending on R & D is strong evidence of the importance of new products to the overall corporate strategy.
- Countless corporations and entire industries owe their meteoric rise to new products: IBM and computers, Polaroid and instant photography, Xerox and xerography, Texas Instruments and microelectronic chips, and the dozens of Japanese and other offshore firms who have penetrated our markets with a myriad of new and often superior products.

Just how important are new products to the corporation? A fairly recent Conference Board study showed that 15 per cent of the current

Exhibit 1.2.   R & D Expenditures by Industry (U.S.)

INDUSTRY	R & D EXPENDITURES AS PERCENTAGE OF SALES
Aircraft and missiles	18.3%
Scientific instruments	8.9%
Electrical equipment	7.8%
Machinery	6.1%
Chemicals	4.2%
Transportation equipment	3.6%
Fabricated metal products	1.3%
Paper and allied products	1.2%
Primary metals	1.1%
All manufacturing industries	3.7%

Statistical Abstract of the United States, 1985.

sales of corporations are based on major new products introduced in the last five years.[3] (This percentage excludes minor new products, modifications, and extensions.) Many firms achieved a much higher performance: the figure ranged from zero to over 50 per cent. As might be expected, firms in stable, commodity-type industries tended to be less dependent on new products than those in specialty businesses, where opportunities for innovation are greater or product life cycles are shorter. Finally, this dependence on new products was similar for manufacturers of industrial goods and manufacturers of consumer goods.

Two-thirds of manufacturers expect their companies to become even more dependent on new products over the next five years (see Exhibit 1.3).[4] The importance accorded new products was particularly strong in industrial-product companies. The evidence is that the new-product game is heating up.

My own studies of industrial-goods firms show that the "dependence" figure of 15 per cent may understate the case. If we consider all new products introduced over the last five years — major and minor, including extensions and significant modifications — then new products represent, on average, 36 per cent of corporate sales.[5] That is, more than one dollar in three in the top line of the profit-and-loss statement comes from products the company didn't have in its portfolio five years ago. In some firms, the figure is as high as 100 per cent — representing a complete turnover of the company's product portfolio in five years.

Exhibit 1.3.   Companies' Expected Dependence on New Products for Future Sales

EXPECTED DEPENDENCE ON NEW PRODUCTS FOR SALES OVER NEXT FIVE YEARS	Total all reporting companies	COMPANIES SELLING PRIMARILY TO	
		Industrial markets	Consumer markets
Higher than now	67%	71%	60%
About the same	25%	21%	31%
Lower than now	8%	8%	9%
	100%	100%	100%

Reprinted with permission from D.S. Hopkins, *New Products Winners and Losers*, report no. 773 (New York: The Conference Board, 1980).

Booz Allen & Hamilton find that managers expect new products to fuel sales and profits in the 1980s at an increasing rate.[6] Indeed, the contribution made by new products to sales growth over the next five years is expected to increase by one-third, while the portion of total company profits generated by new products will increase by 40 per cent.

To support these new-product targets, companies expect to double the number of new products introduced. Exhibit 1.4 shows the dependence on new products as a profit generator. In 1986, new products introduced in the previous five years will contribute 32 per cent of the corporation's profits — an astonishing figure when compared to the 22 per cent dependence cited in 1976. (The industry-by-industry breakdown of these figures in Exhibit 1.4 provides a useful means of comparison for your own firm.) These findings are strong evidence not only of the critical role of new products in corporate prosperity, but also of the increasing magnitude of that role.

Suggestion: If you haven't already done so, conduct a review of the strategic role — past, present, and future — of new products in your company. Key questions include:

- What is your historical level of R & D spending as a percentage of sales? Your proposed future level? How does this compare to your competitors' or industry level? Why is it higher or lower?
- What proportion of your current sales comes from new products introduced in the last five years? What is the projection for the future? (Ask the same question for corporate or divisional profits.)

Exhibit 1.4. Contribution of New Products to Profit (by Industry)

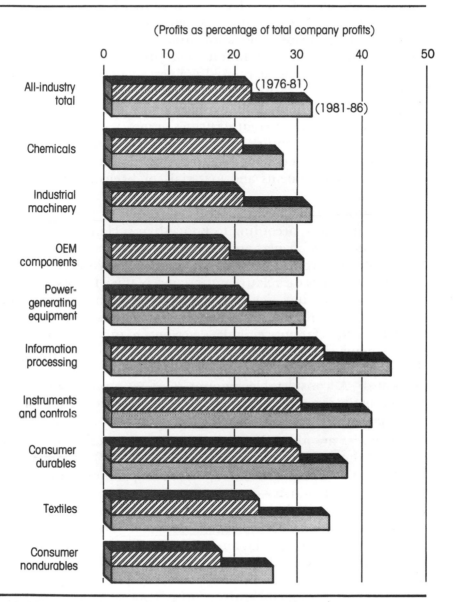

(Profits as percentage of total company profits)

Reprinted with permission from *New Product Management for the 1980s* (New York: Booz Allen & Hamilton, 1982).

What proportion of your growth in sales or profits came from new products (and is expected to come from future new products)?
• Are the answers to the above two questions consistent with each other? Are you investing enough to yield the results you want?

In comparing your company's R & D spending to the industry averages shown in Exhibit 1.2, remember that R & D spending is by no means the sole determinant of new-product performance. In a recent study on firms' innovation strategies, R & D spending (as a percentage of sales) was found to be the strongest predictor of new-product sales (also expressed as a percentage of company sales). This comes as no surprise. But R & D spending explained only 16 per cent of this performance; many other factors also determined performance.[7] Finally, different strategies or means of introducing new products may not require similar levels of R & D spending. Such low-R & D approaches include acquiring technology from others by purchasing components and materials or licensing products and technology.

The strong trend toward more new products is the result of a number of factors, according to Booz Allen & Hamilton.[8] Four key factors cited by managers are technology advances, market requirements, shorter product life cycles, and increased world competition. Exhibit 1.5 shows the relative importance of these factors. Particular obstacles to product development are the high cost of capital, government regulation, and increasing labor costs. Nonetheless, the positive factors far outweigh the deterrents to increased innovation.

Few managers are surprised to hear of the massive resources that industry devotes to product innovation. But how are those resources allocated? A considerable amount goes to capital expenditures, largely in the commercialization phase: capital expenditures amount to 26 per cent of the total expenditures on new products.[9]

Resources are also allocated in varying degrees to the different stages of the new-product process (see Exhibit 1.6). Commercialization — the building of the plant, the tooling of the production line, and the market launch — amounts to about one-quarter of the total cost. But the actual design and development of the product takes up the largest amount, 37 per cent of the total. The initial stages of exploration, screening, and business analysis account for a surprisingly high 21 per cent of the total, almost as much as the commercialization phase. (Note that these early stages involve many projects that never make it through the entire process; hence the high expenditure figure.) Finally, product and market testing amount to 17 per cent of the cost of product innovation. These figures are averages, and differ among companies and industries, but they do provide a useful benchmark for comparison.

Exhibit 1.5.   Impact of External Factors on the Introduction of New Products

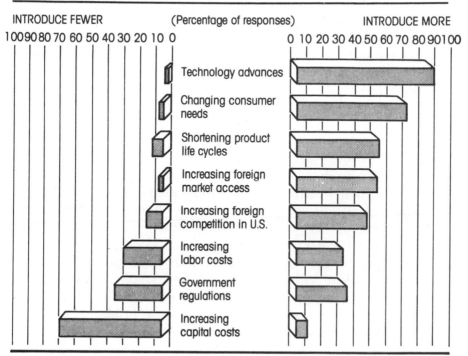

Reprinted with permission from *New Product Management for the 1980s* (New York: Booz Allen & Hamilton, 1982).

Exhibit 1.6.   Relative Expenditures on the Various Stages of the New-Product Process

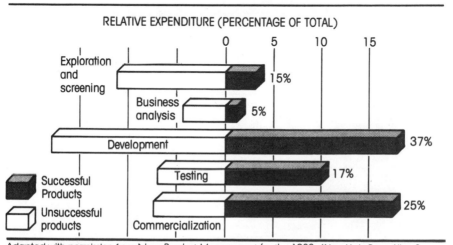

Adapted with permission from *New Product Management for the 1980s* (New York: Booz Allen & Hamilton, 1982).

An Introduction to the Game

In this chapter we have seen that the new-product game can play a critical role in the fortunes of a company. In the next chapter we will look at the reasons new products fail or succeed. Chapter 3 pulls together various companies' experiences and formulates a set of rules or lessons as the new-product game plan is unveiled.

Chapters 4 through 8 analyze the game plan, step by step, from a product idea to a successful product launch. Chapter 4 outlines dozens of ways in which you can double, triple, or even quadruple the number of product ideas you look at each year. Paring this list of ideas down to a set of "best bets" is the topic of chapter 5, wherein ways to heighten the effectiveness of your new-product screening and evaluation decisions are revealed. Chapter 6 demonstrates how to build a strong market orientation into your new-product game plan, and how to arrive at a winning new-product concept.

"No surprises" should be the motto in the new-product game; in chapter 7 the role of testing is outlined. The final play — moving into the market — is the topic of chapter 8, which sets out the details of how to put together a marketing plan for a new product. Finally, in chapter 9 we stand back and look at how our game plan fits into the larger picture — the master strategy for new products.

So read on! Become part of the new-product game and observe the unfolding of the game plan — the blow-by-blow description of how to master the game from idea to launch.

What Separates the Winners from the Losers?

New Products: A High-Risk Endeavor

The new-product game is like a horse race: ten horses leave the gate, and at the end of the race only one is the winner. The gambler tries to pick that one winning horse in ten, but more often than not he places his bet on the wrong horse.

New-product management, however, is far more complex than a horse race. True, the odds of picking a winner at the outset are somewhere in the order of five or ten to one. But the size of the bets is considerably greater — often in the hundreds of thousands and millions of dollars. And unlike the gambler, the new-product manager can't leave the game — he or she must go on placing the bets, year after year, if the company is to succeed.

Finally, the way bets are placed is important. At a racetrack, all bets must be placed before the horses leave the gate. But the new-product manager can and should continue to place bets even as the race proceeds. Indeed, some of the heaviest betting takes place as the "horses" approach the finish line. This ability to place bets during the race improves the odds of picking a winner, but complicates the betting. The gambler who has some type of betting system built into the game plan is more likely to prove successful.

The Odds of Failure

New products that make it to the marketplace face a failure rate of somewhere between 35 per cent and 50 per cent. The figures vary, depending on how one defines "new product" and "failure." Some sources cite the launch failure rate to be as high as 90 per cent. But according to Crawford, who has undertaken perhaps the most thorough review of these often-quoted figures, the true failure rate is about 35 per cent.[1] In a review of the new-product performances of 122 industrial-product firms, the average success rate of fully developed products was 67 per cent.[2] Averages often fail to tell the whole story, however; this success figure varied from a low of zero per cent to a high of 100 per cent.

Other studies point to the difficult times faced by new-product managers: The Conference Board reports a median new-product success rate (defined as success in the marketplace, after launch) of 66 per cent for consumer goods and 64 per cent for industrial products,[3] and Booz Allen & Hamilton report a 65 per cent success rate for new-product launches.[4]

Regardless of whether one faces a 50 or 65 per cent chance of success, the odds of a misfire are still high. Worse, the figures cited above don't include the many new-product projects that are killed before launch and after the expenditure of considerable time and money.

The attrition curve of new-product projects tells the whole story.[5] For every seven new-product concepts, five enter product development, one-and-a-half are launched, and one becomes a commercial success (see Exhibit 2.1). An estimated 46 per cent of all the resources allocated to U.S. product development and commercialization are spent on products that are cancelled or fail to yield an adequate financial return. This is an astounding statistic when one considers the magnitude of human and financial resources devoted to new products.

Exhibit 2.1.   The Attrition Rate of New-Product Projects

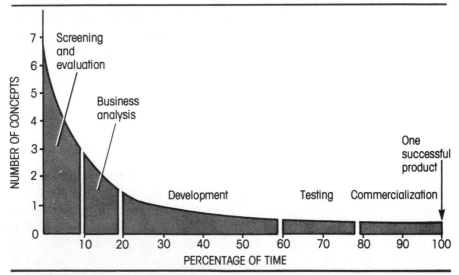

Adapted with permission from *New Product Management for the 1980s* (New York: Booz Allen & Hamilton, 1982).

A review of these disturbing data raises serious questions about firms' and managers' capabilities in the new-product game. It is not surprising to learn that the senior managements of many companies are dissatisfied with corporate new-product performance. In a Conference Board survey done by D.S. Hopkins, only 37 per cent of senior management thought their firms' new-product success rates were "highly acceptable."[6] The remaining 63 per cent thought their success rates were either "disappointing" or "unacceptably low." There was greater dissatisfaction on the part of industrial-goods producers than consumer-products firms.

1. A Minority of Firms Point the Way

Although the average performance of companies is mediocre, some firms feature a stellar new-product performance. As reported in the Conference Board study, for example, 32 per cent of industrial firms and 31 per cent of consumer firms boasted a success rate of new-product launches of 80 per cent or better — an impressive track record. In one of my own studies, one small but enviable group of companies was found to have had a superb new-product record by any standard of success.[7] These companies, which were typically higher-growth, higher-technology, and technologically developing ones, achieved a 72 per cent success rate for new products developed (over 72 per cent for actual launches); 47 per cent of current sales came from new products introduced in the last five years; their new-product programs exceeded their corporate objectives, fared far better than their competitors' programs, were highly profitable, and were rated extremely high in terms of generating corporate sales and profits. Clearly, some firms have succeeded in mastering the new-product game. The strategies of these top-performing firms are revealed in chapter 9.

2. Nowhere but Up

A casual perusal of some firms' new-product records reveals horrifying results. But improvement is possible, and in many companies the timing is right to take action. Some failures are inevitable, of course. No consistently successful firm boasts a 100 per cent batting average, and a company that is serious about new products must expect to have some failures. If the corporate culture and reward/punishment system cannot accommodate occasional product misfires, then management is being unrealistic in its expectations. Remember: the only way to avoid new-product failures altogether is to stop introducing new products!

3. Steady Improvement

The Booz Allen & Hamilton study reports that new-product performance is improving. Ten years ago, for example, 70 per cent of the resources spent by U.S. industry on new products went to the losers; today the figure is less than 50 per cent. The attrition curve also has improved markedly.

4. Reasons for Better Performance

There are good and logical reasons for the top performers' doing so well, and for performance improving over the years. A consistent pattern underlies these success stories, a pattern or game plan considerably different from that of the losers. By taking a close look at what the winners do differently from the losers, and what some companies do differently from ten years ago, much can be learned about the ingredients of a winning game plan. The remainder of this chapter is devoted to identifying these winning factors and ingredients.

Suggestions: What are the failure, success, and "kill" rates in your company? If no one has measured these or developed an attrition curve for new-product projects, perhaps the time is ripe to do this. Companies that develop these data usually find that the type of product greatly affects the attrition or failure rates. For example, a major chemical firm reports attrition rates considerably higher for "blue sky," step-out products than for "close to home" products. You may wish to distinguish between product types; each type of product can be expected to have a different success rate and attrition curve.

Why New Products Fail

An understanding of past failures yields important clues to what should be done differently in the future. Investigations into the reasons that new products fail — a post-mortem of product disasters — helps to identify the pitfalls, obstacles, and deficiencies in new-product management. This knowledge can then be used to take steps to avert or overcome these difficulties in the game plan.

Why do new products fail? Only in the last decade or so has there been hard evidence pinpointing the causes of failure. A number of Conference Board studies into new-product failures have identified marketing variables as the major weaknesses.[8] Exhibit 2.2 shows the main causes of new-product failure.

The major culprit is inadequate market analysis prior to product development: firms simply fail to do their marketing homework before they become seriously involved in the development of the

Exhibit 2.2. Causes of New-Product Failure

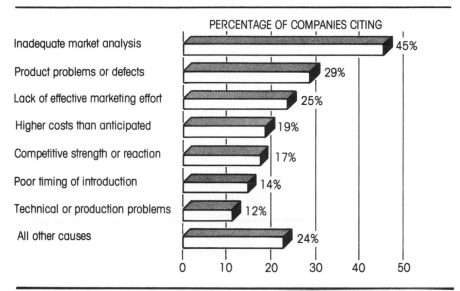

PERCENTAGE OF COMPANIES CITING

Cause	Percentage
Inadequate market analysis	45%
Product problems or defects	29%
Lack of effective marketing effort	25%
Higher costs than anticipated	19%
Competitive strength or reaction	17%
Poor timing of introduction	14%
Technical or production problems	12%
All other causes	24%

Adapted from D.S. Hopkins and E.L. Bailey, "New Product Pressures," *Conference Board Record* 8 (1971): 16–24.

product. This weakness is pointed out in the most recent investigation by the Conference Board.[9] A lack of thoroughness in identifying real needs in the marketplace, or in spotting early signs of competitors taking the offensive, are frequently discovered through postmortem reviews of product failures. Managements confess to a serious misreading of customer needs, too little field testing, or over-optimistic forecasts of market need and acceptance. All too often, the Conference Board study found, managements still fall into the trap described by the vice-president of one industrial firm: "Simply stated, we decided what our marketplace wanted in this new product without really asking the market what its priorities were."

The Gemini case cited in chapter 1 is an excellent illustration. The product was developed by a well-intentioned R & D or engineering group, but in a relative vacuum with little or no marketing involvement or input. Only after the product was on the market did management learn the truth: that there was a serious lack of fit between customer needs and wants and the product's design.

There are other reasons for product failure. Technical deficiencies in the product are the number-two cause of failure. Again, some of these failures can be attributed to a lack of understanding of the customer's needs and a lack of testing of the product in a real-world situation. Technical problems also result from short-cutting some of the key technical and product-testing stages of the process; this suggests the absence of a disciplined game plan and a well-thought-out new-product process.

Bad timing is the third most common cause of product failures. The penalties for moving too slowly or too fast stem not only from technical problems, but also from flawed planning, organization, or control. Higher-than-anticipated costs are another major reason for failure. An insufficient marketing effort at launch, the strength of competitors, and sales force and distribution weaknesses all point to serious weaknesses in a firm's understanding of the new product's marketplace and poor planning and execution of the marketing strategy.

Recommendations in the Conference Board studies call for more and better marketing research, market analysis, and sales forecasting. Traditional marketing research methods — for example, large-sample surveys — are not always the answer, however. Companies have developed many new and more appropriate methods for understanding their markets and customers' needs. Some of these methods are outlined in chapters 6 and 7.

Also high on the list of suggestions for success are more careful product positioning, more effective concept testing, better test marketing, sharper evaluation of new-product projects (including early screening), and better planning and execution of sales and promotional efforts.

My own extensive investigation of new-product failures focused on 114 industrial-product firms.[10] Each company supplied a case history of a new-product failure for an in-depth review. The major reasons for failure were chiefly related to marketing:

- underestimating competitive strength and/or competitive position in the market (36.4 per cent of cases);
- overestimating the number of potential users of the product (20.5 per cent);
- setting the product's price too high (18.2 per cent);
- technical difficulties or deficiencies in the product (20.5 per cent).

Several other reasons were also identified, including selling and distribution weaknesses, bad timing, and actions of competitors.

The same study also pinpointed areas of weakness within the firm and within the new-product game plan itself. Each product failure was reviewed to identify the main activities undertaken as part of the development and launch of the product. With management's help, each step or activity was critically assessed. If the activity was omitted,

Exhibit 2.3. Deficiencies in the New-Product Process

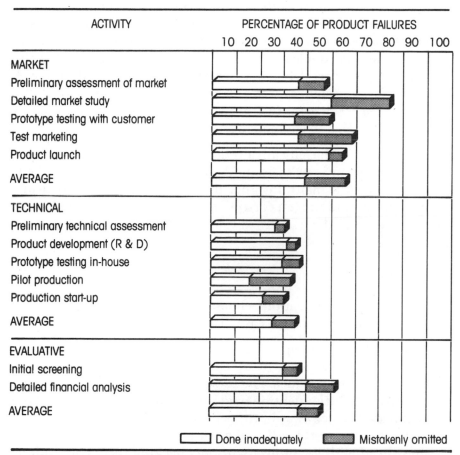

Reprinted with permission of the publisher from "Why Industrial New Products Fail" by Robert G. Cooper, *Industrial Marketing Management* 4: 315–26. Copyright 1975 by Elsevier Science Publishing Co., Inc.

was the omission appropriate? The results are provocative (see Exhibit 2.3). In three-quarters of the failures, the market-research study was either poorly done or omitted altogether. This market-research phase was identified as the most serious deficiency of the entire process. Other market-oriented activities also fared poorly: in more than half of the failures, the test market or trial sell and the market launch were deficient, and in almost half the cases the preliminary assessment of the market and customer-test phase were weak. In contrast, the technical and production activities were more proficiently and consistently undertaken. Though some weaknesses were identified, they occurred much less frequently than in the market-related activities.

Management also identified the resource deficiencies that contributed most to the failure of the products. Consistent with the results above, the marketing areas were singled out as the major weaknesses. A lack of market-research resources — no one in the company capable of doing the needed market studies and analysis — was cited as the number-one resource deficiency in almost two-thirds of the cases. A lack of selling resources was pinpointed by management as the second major weak area. In contrast, other resource areas, such as financial, R & D, and engineering, were much stronger.

Suggestion: If you haven't already done so, undertake a post-mortem review of your company's past new-product failures. The results will provide you with many insights into what ails your own firm's efforts, and should suggest corrective action. The tools to undertake this review are well documented.[11]

Failure Types

A convenient categorization scheme for new-product failures was eventually constructed from case histories.[12]

• *The better mousetrap nobody wanted.* This is the most common type of failure (28 per cent of cases), and typically describes a technology-driven product. The product is conceived and developed internally, with little attention paid to the real needs and wants of the marketplace. Carried away by the belief that they have a better mousetrap and that the world will beat a path to their door, management pushes ahead with the project without checking their assumptions about the market and customers' needs. Only after launch does management discover that no real need for the product, as designed, exists. The Gemini 2000 fits perfectly into this category.

• *The "me too" product meets a competitive brick wall.* This type of failure (representing 24 per cent of cases) is the opposite extreme. The project is often initiated when a successful competitive product is observed. The cry goes out: "We have to have one too!" The strategy is to develop a product remarkably similar to the competitor's in the mistaken belief that simply being in the market will bring a "fair share" of sales. Once the product hits the market, sales fall below expectations. Management suddenly discovers that their offering is identical in features and price to that of an entrenched competitor. The customer has no reason to switch. Management learns that there's no such thing as automatically gaining "our fair share of the market"; it must be earned. And merely "being there" is not enough; the product must be there *and* be better.

These two types of failure scenarios together account for more than half of all new-product failures. There are other less common scenarios, however.

• *Competitive one-upmanship.* Competitors may deliberately set out to upset or destroy a new product's success (13 per cent of cases). For example, a competitor may cut prices just prior to your launch; or it may launch promotions, deals, or a sales push, again just prior to your own launch. Various tactics may be used to destroy your test market or trial sell, such as disrupting your store display, price cutting, or heavy promotions. Finally, competitive product announcements may appear which are designed to take the wind out of your sails. Often, the competitive product is nonexistent or years away from the market; but the announcement is timed to hurt your new product's launch.

• *Technical dog.* The product simply doesn't work, or falls short of performance requirements (15 per cent of cases). A memorable example is the Adam home computer, which was rushed to the market just before Christmas in 1983. Unfortunately, there were technical bugs in the product, and reports soon began to appear in the media about the computer's shortcomings. The Adam was a well-designed home computer, and gave good value for the money. But because it was plagued by technical problems (which were subsequently solved) the product was finally removed from the market.

• *Price crunch.* The new product's price is too high (13 per cent of cases). In some instances, competitors drop their own prices when confronted with a new product on the market. More often, however, the pricing is a consequence of a misreading of the market: too much is built into the product. As one manager put it, "The market wanted a Ford, and we gave them a Cadillac."

• *Environmental ignorance.* Just about everything that can go wrong does go wrong (7 per cent of cases). The product is wrong for the customer; competitors introduce similar products; the selling effort is inadequate and incorrectly targeted; or the products run afoul of government regulations. The disaster results from a complete misreading of the external environment — customers, competitors, and government.

The Key to New-Product Success

Identifying what makes a new product a success is considerably more difficult than pinpointing reasons for failure. A number of experts have attempted to identify the factors in new-product successes. One of the earliest investigations, by Myers and Marquis, looked at 567 successful product innovations, and concluded that most are market-pull projects; only 21 per cent are technology-push.[13] Correct identification of an existing demand is the common ingredient among these success stories.

Internal sources of information are also critical to the innovation process, pointing to the need to foster interaction among departments involved in the new-product process. External information obtained via nonstructured channels is also important. Myers and Marquis were among the first to recognize that a "game plan" exists — that some firms had in place a logical flow of activities, from idea to launch. A simplistic five-step model was proposed as a result of studying these 567 successes.

The General Electric Laboratories provided another setting for an investigation of product successes.[14] Roberts and Burke looked closely at six successful GE products, and concluded that both technological and market variables decide their fates. The successful products had several things in common:

- Market needs were recognized, and R & D was targeted at satisfying those needs.
- When a technological success did not meet a specific market need, the product was adapted to suit an identified need.
- Research managers communicated the possibility of a technological breakthrough clearly to other departments, which facilitated the identification of a market need.
- Communication existed between engineers and scientists and other involved (operating) departments.

Thus, in spite of the fact that the products were all moderate- to high-technology products, emphasis was placed on market needs and market-need identification.

In one of my investigations of three significant high-technology new industrial products — a new milk-packaging system by DuPont of Canada, a new telephone by Northern Telecom, and a new jet engine by United Technologies — much was learned about what went into successful product development.[15] The one common thread in these developments was a strong commitment and orientation to the marketplace. In all three cases, there was extensive and careful analysis of the marketplace; in fact, in two instances, eleven separate market studies were undertaken. The market studies dealt not only with the more obvious issues, such as market potential and size, but also with the nature of customers' needs, the benefits they desired in a product, and the design requirements for a winning product.

The three cases dramatically demonstrated the importance of marrying technological prowess to a market orientation, and the need to undertake one's marketing homework before product development begins. The result in all three cases was the development and launch of a new product that not only was technologically superior and unique, but also met customer needs and delivered unique benefits to end users far better than anything else on the market.

A second common facet of the three winning products was the logical and stepwise flow of activities as the projects moved from idea to launch. Although the three products represented different industries, the flowcharts that portrayed each firm's new-product process bore a striking similarity to one another. It was almost as though the three firms had adopted the same game plan.

A final discovery was the extent of the interaction between functional groups within the firms. The process, from idea to launch, "crisscrossed" back and forth between marketing and technical groups. Although there was a "product champion" or project leader in each case, the effort was by no means a one-person or one-department show. An interdisciplinary approach, with strong interaction between functional groups, was evident in all three cases.

The Winners Versus the Losers

What separates the winners from the losers? Myers and Marquis discovered that three-quarters of all successful products were market-derived. The naïve observer might conclude that market-derived projects are the key to success, rush back to his or her firm, and begin a program of developing market-derived projects. Remember, however, that it is conceivable that three-quarters of all *failures* also are market-derived. In order to find the keys to success, we must compare the factors that distinguish successful and unsuccessful products.

The U.K. Experience

A British study, Project SAPPHO, is considered a classic: it was one of the first of success-versus-failure studies, and identified a pattern of differences between a paired sample of 43 successful and unsuccessful innovations.[16] Of the 122 variables measured, 41 were found to discriminate between successes and failures, and five underlying factors or themes distinguished the winners.

The most important success factor was an understanding of users' needs; more than any single element of the success formula, the ability to identify and understand user needs, wants, preferences, and benefits was the critical deciding variable. Other important factors were efficiency of development; certain characteristics of management and managers; effectiveness of communication, both internal and external; and magnitude of the marketing effort at launch.

SAPPHO in Other Countries

A smaller study of the Hungarian electronics industry yielded comparable results.[17] In spite of the political, cultural, and economic differences between the two countries, a similar set of success factors was identified:

- market-need satisfaction;
- effective communication (internal and external);
- efficient product development;
- a strong market orientation; and
- the roles of key individuals.

The SAPPHO researchers recently reported the results of a five-country study of innovation in the textile-machinery industry.[18] High-performance firms shared certain characteristics:

- they had superior marketing capability and frequent customer contact;
- they understood users' needs and were able to assess whether these needs could be filled economically;
- they carefully matched specific sales strategies to market requirements.

It was found that firms that employed qualified scientists and engineers were more able to produce successful breakthroughs; and more radical innovations stemmed from those firms with technically qualified chief executives.

The U.S. Experience

Fifty-four significant facilitators for success were identified in Rubenstein's investigation of U.S. new-product success and failure pairs.[19] No single factor of success or failure could be isolated, however. Further, the researchers note that one person's facilitator can be someone else's barrier. The more important facilitators included:

- the existence of a "product champion";
- marketing factors, such as need recognition;
- strong internal communication (between functional areas in the firm);
- superior techniques for data gathering, analysis, and decision making; and
- planned approaches to venture management.

A recent study, the Stanford Innovation Project, yielded results consistent with the SAPPHO and NewProd studies (described below). By comparing a large number of new-product successes and failures, Maidique and Zirger concluded that new product success is likely to be greater when:

- the firm developing the new product has an in-depth understanding of the customers and the marketplace;
- the marketplace is attractive, as measured by its rate of growth, size, and competitiveness;
- the firm is proficient in marketing and commits a significant amount of its resources to selling and promoting the product;
- the product is unique and is perceived by the customer to give good value for money;
- the project is supported by management throughout its development and launch;
- the development process is well planned and coordinated; and
- the markets and technologies of the new product benefit significantly from the existing strengths of the firm.[20]

Project NewProd

Project NewProd probed the key factors that separated successful new products from failures.[21] Almost 200 products from 100 industrial-product firms were studied. Eighty characteristics of each project were measured, including the market, the technology employed, the activities that were undertaken as part of the innovation process, and the nature of information gathered during the process. These characteristics were then related to the project outcomes — commercial success or failure — in order to identify the deciding factors. A commercial success was defined as a product whose financial return exceeded the minimum acceptable return for that type of investment.

1. Product Superiority

The most important single success factor was having a superior new product that delivered significant and unique benefits to the end user. The odds of success with a unique, superior product were over 80 per cent; in contrast, the "me too" or "ho hum" products achieved a success rate of only 28 per cent. Superior products were three times more likely to succeed than the "me too" products. Exhibit 2.4 shows the dramatic impact of product superiority on new-product success.

Exhibit 2.4. Impact of Product Superiority on Success

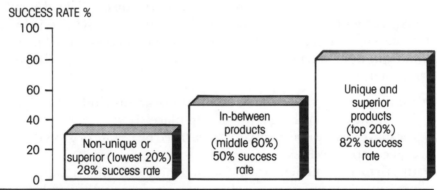

Reprinted with permission from R.G. Cooper, "The Myth of the Better Mousetrap: What Makes a New Product a Success," *Business Quarterly* 46 (Spring 1981): 69–81.

What are the ingredients of a superior product? They offer unique features to the customer, meet customer needs better than competitive products, perform a unique task for the customer, lower customers' costs, and are of higher quality than competing products. The relative effects of these factors are shown in Exhibit 2.5. While technological prowess was obviously important to the development of the superior products, an understanding of the marketplace was critical to their successful design and development. Somewhat disconcertingly, we found that the great majority of new products were anything but "superior products." The sad fact is that the average new product studied scored only five or six out of ten on the list of key factors in a superior product.

2. A Strong Market Orientation

The second key to success was a strong market and marketing orientation; this clearly parallels the first success factor. Again, the impact of the factor was dramatic: strongly market-oriented projects were

Exhibit 2.5. The Ingredients of a Superior Product

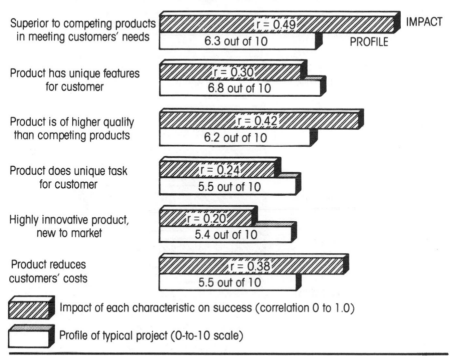

Superior to competing products in meeting customers' needs — $r = 0.49$ — 6.3 out of 10 — IMPACT PROFILE

Product has unique features for customer — $r = 0.30$ — 6.8 out of 10

Product is of higher quality than competing products — $r = 0.42$ — 6.2 out of 10

Product does unique task for customer — $r = 0.24$ — 5.5 out of 10

Highly innovative product, new to market — $r = 0.20$ — 5.4 out of 10

Product reduces customers' costs — $r = 0.38$ — 5.5 out of 10

Impact of each characteristic on success (correlation 0 to 1.0)

Profile of typical project (0-to-10 scale)

Reprinted with permission from R.G. Cooper, "The Myth of the Better Mousetrap: What Makes a New Product a Success," *Business Quarterly* 46 (Spring 1981): 69–81.

successful 79 per cent of the time, while those rated the weakest in terms of market orientation scored only a 28 per cent success rate.

Market-oriented projects were characterized by extensive market information gathering: preliminary market assessment; detailed market studies or marketing research; concept tests in the market-place; prototype trials; and even test marketing or trial selling (see Exhibit 2.6). Market information was very complete: there was a solid understanding of the customers' needs, wants, and preferences; of the customer's buying behavior and price sensitivity; of the size and trends of the market; and of the competitive situation. Finally, the market launch was well planned, well targeted, proficiently executed, and backed by appropriate resources.

3. Technological Fit and Proficiency

The final major key to success was competent technological and production activity and a high degree of fit or synergy between the technological needs of the project and the resource base of the com-

Exhibit 2.6.   Ingredients of a Strong Market and Marketing Orientation

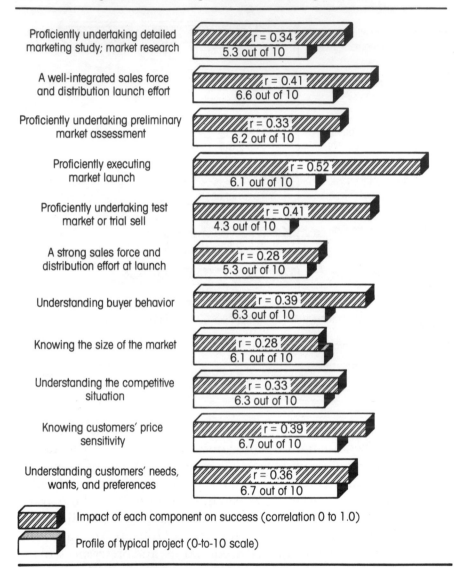

Proficiently undertaking detailed marketing study; market research
r = 0.34
5.3 out of 10

A well-integrated sales force and distribution launch effort
r = 0.41
6.6 out of 10

Proficiently undertaking preliminary market assessment
r = 0.33
6.2 out of 10

Proficiently executing market launch
r = 0.52
6.1 out of 10

Proficiently undertaking test market or trial sell
r = 0.41
4.3 out of 10

A strong sales force and distribution effort at launch
r = 0.28
5.3 out of 10

Understanding buyer behavior
r = 0.39
6.3 out of 10

Knowing the size of the market
r = 0.28
6.1 out of 10

Understanding the competitive situation
r = 0.33
6.3 out of 10

Knowing customers' price sensitivity
r = 0.39
6.7 out of 10

Understanding customers' needs, wants, and preferences
r = 0.36
6.7 out of 10

Impact of each component on success (correlation 0 to 1.0)

Profile of typical project (0-to-10 scale)

Reprinted with permission from R.G. Cooper, "The Myth of the Better Mousetrap: What Makes a New Product a Success," *Business Quarterly* 46 (Spring 1981): 69–81.

pany. Knowing your technological strengths and selecting new product projects that build on those strengths are clear success factors. This finding points strongly to the desirability of a "stick to the knitting" strategy for new products, at least in terms of technology and production.

Other Factors

The NewProd study found that many other variables influence new product success, although not as strongly as the three factors cited above. Most of the success factors are amenable to management action — that is, are controllable — and success is not nearly as dependent on noncontrollable or situational factors (for example, the market, the nature of the company, or the type of projects) as might have been expected.

Finally, many weaknesses were identified in firms' new-product processes. Many companies lacked an effective new-project screening procedure or scheme. There were far too many "me too" products (essentially copies of competitors' products), and products which, while new and different and often technologically innovative, offered no new benefits to customers. In many cases, vital market information was absent during the actual development of the product. Critical market investigations were often omitted from the process altogether, usually for very poor reasons, and the market launch itself was frequently poorly planned and executed.

Suggestion: Have you discovered what makes for a successful new product in your company? More important, what is it that separates the winners from the losers? A logical way to begin the design of a game plan is with a review of past experiences in order to identify the "necessary" and "desirable" features of a winning product. The exercise involves identifying a reasonable sample of past launches and comparing and contrasting the winners and the losers.

New Corporate Practices

A number of relatively new corporate practices have contributed to the improvements in the performance of many firms' new-product programs.[22] These positive practices were not commonly seen even a decade ago:

- Many companies now use a formal new-product process — a stepwise game plan from idea to launch. Moreover, companies with the most successful new products are more likely to have had a formal new-product process in place for a longer period of time.
- In this game plan, and preceding the idea phase, is a relatively new step — formulating the new-product strategy — developing a master plan for product innovation.
- This more sophisticated new-product process has dramatically reduced the attrition rate of new-product ideas and projects. Com-

panies with a strong record of new-product introductions consider fewer ideas per successful launch, and do more and better screening of ideas.

- Management attention and resources are increasingly devoted to the early steps of the new-product process. Ten years ago, roughly one-half of all product expenditures was spent on the commercialization phase; today this amount has shrunk to one-quarter, while the front-end activities — exploration, screening, and business analysis — have risen from 10 per cent to 21 per cent of the total tab. A comparison of product development in the United States with that in Japan, however, reveals that the Japanese invest more time and give more attention to these early steps. (Review the breakdown of expenditures shown in Exhibit 1.5. Note that these expenditures represent actual rather than desired practice.)
- More companies are actually measuring the performance of their new-product programs (two-thirds of U.S. firms do so). Commonly used measurement criteria are profit contribution, return on investment, and sales volume from new products.
- Most companies (over one-half) use more than one type of organizational structure to manage and undertake new-product projects. Most tie the choice of structure used to product-specific requirements. Many firms are now using free-standing or autonomous units, such as interdisciplinary teams, separate new-product departments, and venture groups. Most also continue to use the traditional functionally based units that are part of an existing department, such as R & D, marketing, engineering, or planning.

Suggestion: Conduct a "new-product practices" audit in your company. Consider these questions:

- Does your firm have a formal new-product process or game plan in place? Do you follow it?
- Do you place sufficient emphasis on the up-front or predevelopment activities — exploration, project selection, and business analysis?
- Does your firm measure the results of its new-product program? Do you know what those results are?
- How do you organize for new products? Is the new-products activity housed in a traditional functional area such as R & D, Marketing, or Engineering? Or have you considered an alternative structure to manage this vital function?

Winners and Losers

While luck, no doubt, plays a role in deciding the outcome of the new-product game, a closer look at the winners and losers reveals distinct patterns. Success is clearly more than good luck. Clear and consistent factors — management practices, the game plan, and the types of products and markets selected — separate winners from losers.

This chapter has provided solid evidence of what it takes to be a winner. In the next chapter, we pull together this evidence, draw out its implications in the management of new products, and shape it into a winning game plan.

The New-Product Game Plan

What can we learn from investigations into new-product performance? A number of underlying themes and recurring messages begin to emerge as one examines the experiences of companies. Let us now consider the more evident lessons and reflect on how we can benefit from each lesson and translate each into an operational facet of our game plan.

Lesson 1. New-product development and commercialization is a high-risk but vital endeavor of the modern corporation.

New-product development is a crucial source of corporate growth and prosperity. New products probably will play an even more important role in the years ahead. Business history abounds with stories of corporate fortunes that were made and lost on new products.

- About 35 per cent of the current sales of corporations come from products that weren't in their portfolios five years ago. The proportion of sales derived from new products grew by one-third between 1981 and 1985.
- New products accounted for about 22 per cent of corporate profits in 1981. This figure will grow to 32 per cent by the end of 1986, an increase of 40 per cent.
- By a margin of eight to one, senior executives believe that their corporations will be even more dependent on new products as a source of growth in sales volume in the years ahead.

New-product development is an extremely risky endeavor. The payoffs from a successful new product or innovation program are obvious, but so are the potential losses: U.S. industry spends an estimated 2.6 per cent of sales on R & D. Expenditures on all facets of new products — analysis, development, and commercialization — are likely much higher than that. The attrition and failure rates of new products tell a sobering story:

- Of every seven new product ideas, only one becomes a commercial success.
- Of every four new products that are fully developed, only one becomes a commercial success.
- One out of every three new products actually launched into the marketplace fails commercially.
- Almost half the resources that U.S. industry devotes to the development and commercialization of new products is spent on products that fail or are cancelled.

Given the high potential payoffs and the significant potential losses, every effort must be made to improve the management of the game. The odds of failure are simply too high, and the game is too important to lose.

Lesson 2. There are no easy answers to what makes a new product a success.

Success does not depend on one or even a few factors. Nor can success be guaranteed by carrying out a handful of activities particularly well. Success depends on doing many things well; a single miscue can spell disaster.

In watching a team sport, how many times have you heard the sportscaster comment: "The team played a great game for 58 minutes. Unfortunately, they had a few lapses in the last minutes of play, and lost the game." A lapse of several seconds, a player out of position, or a mistake in strategy and the game is over. The new-product game is much the same. The new-product manager faces the complex task of managing a highly uncertain endeavor in which many tasks must be performed perfectly, and in which a single error can be fatal. The manager, like the team coach, must rely on a detailed game plan. A carefully developed blueprint for action is crucial to ensuring that no vital activity or step is overlooked. Discipline is essential in the new-product game.

Lesson 3. Companies that follow a new-product game plan do better.

Those firms that have adopted a stepwise approach to product innovation perform better, and those that have had the approach in place for a longer time do better yet. As one vice-president put it, "The multistep new-product process is an essential ingredient in successful new-product development."

This conclusion parallels the findings in a study of three highly successful new products.[1] In all three cases, management was convinced that a disciplined approach — following a logical sequence of steps from idea to launch — left much less to chance. By following the game plan, steps that are often handled poorly or omitted altogether are more likely to receive the attention they should. If your firm does not have such a game plan in place, the design and implementation of a plan should become a top-priority task. Later in this chapter we will look at an outline of a typical game plan — a step-by-step procedure for turning a new-product idea into a winning new product in the marketplace.

Lesson 4. New-product success is amenable to management action.

Most researchers agree that the actions of the people involved in new-product projects largely decide the fates of the ventures. Product innovators are not victims of circumstance or prisoners of their environment. The outcome of a venture depends not so much on the nature of the market, the competitive situation, or the nature of the project, but rather on what people do about it. The role of key players (for example, the product champion) and the impact of certain pivotal activities (for example, market research, test marketing, and product launch) underscore this point. R. Rothwell, the principal investigator behind the project SAPPHO studies, notes that "while chance and uncertainty can upset even the best-laid schemes, responsibility for the success or failure of innovations ultimately rests firmly in the hands of the innovating companies' own management."[2]

In contrast, adherents of one school of thought argue that the outcome of a new-product project depends mostly on the circumstances — on the market, the competition, and other environmental or outside factors. These "product fatalists" express their message in such sayings as "One has to be in the right market at the right time" and "Even a blind man can get rich in a gold mine by swinging a pick-ax." The assumption is that product outcomes are largely decided by extraneous factors; that there is little that management can do to turn a mediocre project into a star; and that new-product outcomes are largely pre-ordained.

A logical outgrowth of such fatalistic thinking is a preoccupation with project selection. In some firms there is far more focus on the question, "Is this the right project?" than on the question, "How do we go about turning this project into a winner?" Much time and effort goes into creating reports or studies in one form or another and sophisticated financial analysis directed at project justification, rather than getting on with the job of creating a winning new product. This attitude is sometimes referred to as "paralysis by analysis." If the same effort was devoted to moving ahead with the project, the company probably would have a successful product in the market.

Project selection or evaluation is not unimportant, but the actions of people as they design and execute the various stages of the project have more to do with deciding the product's ultimate fate than the uncontrollable factors that usually underlie the selection decision, especially near the beginning of a project.

A certain amount of time and effort must be devoted to the "GO/KILL" decisions inherent in the process. But once a GO decision is made, get on with the job of "doing the project." This means the design and execution of a series of activities that moves the project through its various stages in the game plan. As one manager put it, "We first decide what river we're paddling in; then, having decided, we get on with the job of figuring out how to paddle, and how to paddle well!"

Lesson 5. A strong market orientation is needed in new-product development.

Need recognition and market research are identified as the keys to success in virtually every recent study of new-product performance. Accurate market information should play a pivotal role in shaping the product's design and in its eventual launch strategy. Most successful new products are market-derived: the product idea is a result of the identification of an unsatisfied customer need. In the case of successful technology-push products, innovators determine the existence of a market need before proceeding; they then define users' needs and attempt to ensure that the design of the new product satisfies those needs.[3]

It is clear that many firms are deficient in their market orientation. Too little market research is undertaken too late in the process. Many firms confess to a lack of ability to do the kind of market studies so important to a successful new product. In fact, market studies are often omitted altogether. It bears repeating that inadequate market research remains the number-one reason for new-product failure.

The ideal new-product process would balance technical and engineering research with extensive marketing research. Market information should be integrated into every stage of the process, not just as an afterthought at the time of the launch phase. Whether the project is market-pull or technology-push, market information should be used not only in the evaluation or GO/KILL decisions, but in product-design, engineering, and development activities.

Lesson 6. The nature of the product is central to its success.

The central role of the product itself — its design, features, advantages, and benefits — in achieving success should come as no surprise. A number of studies, however, point out that "tired" products and "me too" offerings are the rule rather than the exception in many firms' new-product efforts. But merely having an innovative product is not sufficient. The product must be unique and superior in the eyes of the customer, not just in the opinion of the R & D department. The fact that the product may represent a technological breakthrough or utilizes a new material may excite the R & D group but leave the customer cold. Unless the product delivers unique benefits to the customer, its technical niceties are irrelevant. Remember that a product is simply a "bundle of benefits" for the customer. When determining whether or not your product is truly superior, look at it from the customer's perspective: does it deliver unique benefits to the user?

This desire to offer a "better product" in terms of meeting customer needs parallels the need to be strongly market-oriented. The firm must possess the technological and production expertise necessary to develop and produce the "superior" product: The product may be "better" because it uses new technology, or is better designed, or is produced at a lower cost than its competitors. If the product is to deliver significant advantages and benefits to the user, however, a clear understanding of the customer's needs, wants, preferences, and choice criteria is essential before serious product development begins. The value of a product is in the eye of the beholder — the customer.

Lesson 7. More homework must be done before product design and development are undertaken.

The steps that precede the actual design and development of the product — screening, market studies, feasibility analysis, and business analysis — are crucial to the success of a new product. Errors and omissions in these activities can and often do spell disaster for the project.

Studies of new-product failures show that weaknesses in the up-front activities seriously compromised the projects. A lack of market assessment, a failure to arrive at a winning product concept, inadequate project screening and evaluation, and poor definition of the target market were familiar themes. Consider the Gemini 2000 case described earlier. Virtually no up-front activities were undertaken: project evaluation was weak; no market studies were done; there was only a vague definition of the target market; and the product concept and specifications were only vaguely defined. It is unfair to blame the product-development team — the engineers and designers — for creating a "bad" product design: they were working in a vacuum.

Today, many companies are investing more in the up-front steps. "Successful companies conduct more analysis early in the process and focus their idea and concept generation. And they conduct more rigorous screening and evaluation of the ideas generated. ... The Japanese invest even more time and give more attention to these early steps in the process."[4] The proportion of spending on homework activities has more than doubled in the last decade or so, and now represents 21 per cent of total new-product expenditures.

Why the relatively sudden emphasis on these up-front activities? First, project screening and evaluation — the GO/KILL decisions — are obviously crucial: too lax a screening process leads to many mediocre projects going into development and the resulting misallocation of scarce development resources. Second, it is in these early stages that the product and project must be defined: the target market must be pinned down and understood; the product concept and the product's positioning in the marketplace must be accurately spelled out; and the product's design specifications must be agreed upon. All this must be done prior to moving into full-scale product development. Otherwise, the result is likely to be an expensive and inefficient product-development effort that lacks targets, objectives, and direction.

Lesson 8. Better, more consistent, and more systematic product evaluation is required.

Successfully planning and executing the steps of the new-product process is half the battle; picking winning projects in the first place is the other half. The first evaluation, project or idea screening, is still handled in an ad hoc, informal manner in many firms. That's a polite way of saying that no screening at all takes place; projects seem to slip into the process, almost by osmosis. If any screening is done, the procedures and criteria used to evaluate the projects are inconsistent. Subsequent evaluations also are poorly done; in many cases, inappropriate techniques are used (for example, rigorous financial analysis is used prematurely, when estimates of expected sales, costs, and margins are likely to be pure guesswork); critical information essential to a thorough evaluation is missing (for example, realistic estimates of market size and expected sales); or negative information and evaluations are ignored altogether.

Consider the case of Procter & Gamble's Pringle's potato chips. This project failed or was marginal on most of the evaluations (preference tests, pre-test market, and test market), yet still had sufficient momentum to slip by despite the evaluations. The product was eventually launched, but proved to be a financial dud for P & G. Today, after valiant attempts to remedy the situation, the product continues to limp along. (This is not a criticism of P & G's evaluation procedures; the company has in place some of the toughest and most thorough project-evaluation procedures — a model for other firms to emulate. But even there, some losers slip through.)

Many studies recommend the use of sharper and more consistent screening and evaluation procedures and criteria. Some firms are paying attention. There is a renewed interest in establishing screening mechanisms, ranging from simple checklists to more sophisticated project-scoring models (checklists with weighted questions). Studying historical reviews of what makes a new product a winner is useful in developing such checklists. Some firms have set up screening committees to scrutinize projects; others use a panel of outside judges in order to obtain a more objective evaluation. Finally, better market information, derived from market studies, helps to heighten the effectiveness of the evaluation. In chapter 5 we take a much closer look at the question of project screening and evaluation — how to pick a winner.

Lesson 9. A well-conceived, properly executed launch is vital to success.

A strong marketing effort, a well-targeted selling approach, and effective after-sales service contribute significantly to the successful launching of a new product. But a well-integrated and properly targeted launch effort does not happen by chance; it is the result of a fine-tuned marketing plan properly carried out.

The marketing-planning process — moving from marketing objectives to strategy and marketing programs — is a complex process. Entire books have been devoted to the subject. Somehow, this marketing-planning process must be built into the ideal new-product game plan. For example, the selection of a target market, one of the first steps in developing a marketing plan, logically should precede product design and development. How do customers buy? What are their choice criteria? What are their sources of information on similar products? What are the competing products and how are they priced? Again, we see a need for market research, but this time it should be directed at providing information essential to the design of a launch plan.

• The design of a marketing plan is an integral part of the new-product process. An effective launch depends on a well-conceived launch plan.

- The marketing-planning process must begin early in the new-product process. It should not be left as an afterthought, to be undertaken as the product nears the commercialization phase; or, as one manager says, "When the product's rolling down the production line, that's when our sales and marketing people become involved." The marketing-planning activities must begin *prior to* product design and development and continue in parallel with product development, right through to the launch phase.

- A marketing plan, like any strategic plan, is only as good as the intelligence on which it is based. Market studies designed to yield information crucial to marketing planning must be built into the new-product game plan.

Lesson 10. Organizational structures that provide for multi-functional inputs and internal communication and coordination foster successful new products.

Many of the investigations into product successes cited interfaces between R & D and marketing; coordination among key internal groups; multidisciplinary inputs to the new-product process; and novel organizational structures designed to maximize interaction and coordination among the various players. Except for the simplest of products and projects — line extensions, product updates, and the like — product innovation cannot be a one-department show. The winning game plan involves a number of varied activities that fall into different functional areas within the firm: market research, R & D, sales, engineering, industrial design, production, purchasing, and advertising, to name a few.

How does one design a game plan that integrates these many activities and multiple inputs? By developing a systematic approach to product innovation that cuts across functional boundaries and forces the involvement of various groups. Equally important, what type of organizational structure will bring many players from different walks of life in the corporation together in an integrated effort? In short, how do we take a diverse group of players and turn them into a team?

It's clear that the traditional functional area organizational structure does not suit many of the needs of product innovation. Some corporations are turning to alternative organizational structures to solve the problem. According to one manager, "The more innovative the project, the more innovative the organizational structure must be to accommodate the needs of the project." Many firms are now using organizational structures such as new-product departments, venture teams (teams that work full-time on a specific new-product project and function relatively independently of the organization), and other unorthodox means of solving the multidisciplinary-team-building problem. In designing a winning game plan, we must be constantly aware of the need to build in multidisciplinary inputs.

Suggestion: Review the ten lessons set out above. List the specific steps or actions that your company has taken in its approach to new products. In the context of each lesson, ask "What is the company doing about this? What should it be doing?"

Managing Risk

The management of new products is the management of risk. Most of the ten lessons set out above dealt with ways of reducing risk. Total risk avoidance in new-product development is impossible, unless a company decides to eschew all innovation — and face a slow death.

Most of us know what is meant by the phrase "a risky situation." From a new-product manager's standpoint, a high-risk situation is one in which large amounts of money or resources are at stake, and in which the outcome is uncertain. Exhibit 3.1 illustrates the components of risk.

Exhibit 3.1. Components of Risk in a Decision Situation

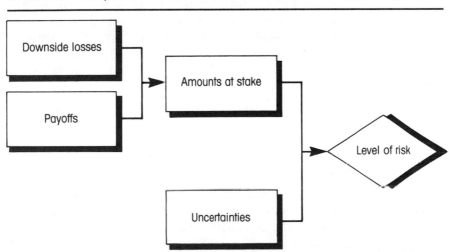

Adapted with permission from R.G. Cooper and R.A. More, "Modular Risk Management: An Applied Example," *R & D Management* 9 (Feb. 1979): 93–99.

A Life-or-Death Gamble

Imagine for a moment that you are facing the gamble of your life. You've been invited to a millionaire's ranch for a weekend. Last night, you played poker and lost more money than you care to admit — around $20,000. All of the other players are enormously wealthy oil and cattle barons. Tonight they've given you the opportunity to cash in. Each of the other ten players antes into the pot $1 million. That's $10 million in thousand-dollar bills stacked up on the table in front of you — more money than you're ever likely to see again in your lifetime. But the gamble has its downside risk. One of the players takes out his six-shooter, removes all the bullets, and in full view of everyone places one live bullet into the gun. He then spins the chamber. For $10 million, you are asked to point the gun to your head and pull the trigger. You have one chance in six of dying. Will you take the gamble?

This situation exhibits the key elements of risk — a great deal at stake (the $10 million and your life), and a high degree of uncertainty. True, there is only one chance in six that the downside will occur; but probabilities don't really matter after the trigger is pulled.

Reduce the Stakes

This hypothetical gamble presents an unacceptable risk level. How can the risk be reduced? One step is to lessen the amounts at stake. For example, use a blank bullet and earmuffs to deaden the noise, and point the gun not at your head, but at your foot. The potential downside loss is now merely humiliation in front of a group of friends. Correspondingly, instead of anteing in $1 million dollars each, every player now puts one dollar each into the pot. Will you still take the gamble? Most people would reply, "Who cares?" There is no longer enough at stake to make the gamble worthwhile or even interesting. The risk is so low that the decision becomes trivial.

Some Gambling Rules

Rule number one in risk management is: if the uncertainties are high, keep the amounts at stake low. Rule number two is: as the uncertainties decrease, the amounts at stake can be increased. These rules ensure that risk is kept to a minimum.

There is another way in which risk can be managed in our hypothetical example. The pot remains at $10 million, a live bullet is used, and the gun must be aimed at your head. But this time your opponent, in plain view, marks the chamber containing the bullet with nail polish. He spins the chamber and asks you to reach into your pocket and give him $10,000 in return for a look at the gun to see where the live bullet ended up. Then you decide whether or not you're still willing to take the gamble.

Most of us would consider this a "good gamble" (assuming we had the $10,000) — one with an acceptable risk level. A relatively small amount of cash buys a look at the gun and the location of the bullet. Having paid for the look, you then make your second decision: are you still in the game?

The risk has been reduced by changing one "all or nothing" decision into a two-stage decision. Your ability to purchase information was instrumental in minimizing the risk: the uncertainty of the situation was reduced and an opportunity to withdraw from the wager was created.

This second version of our hypothetical gamble illustrates three more rules of managing risk. Rule number three is: incrementalize the decision process: break an all-or-nothing decision into a series of decisions. Rule number four is: buy information to reduce uncertainty. And rule number five: provide for bail-out points; give yourself an opportunity to get out of the game.

Risk in New-Product Management

The five rules of risk management apply directly to the new-product game. Near the beginning of a project the amounts at stake usually are low, and the uncertainty of outcome is very high. As the project progresses the amounts at stake begin to increase (see Exhibit 3.2). If risk is to be managed successfully, the uncertainty of outcome must be reduced as the stakes increase, and the stakes must not be allowed to increase unless the uncertainties are reduced.

Unfortunately, in many new-product projects the amounts at stake increase as the project progresses (see Exhibit 3.3) while the uncertainties remain fairly high. By the end of the project, as the launch nears, management is no more sure about the commercial outcome of the venture than it was on day one of the project. The amounts at stake have increased, uncertainty remains great, and the risk level is unacceptably high.

For every thousand-dollar increase in the amounts at stake, the undertainty curve must be reduced by an equivalent amount. To do otherwise is to let risk get out of hand. In short, every expenditure in

Exhibit 3.2. Relationship Between Uncertainties and Amounts at Stake

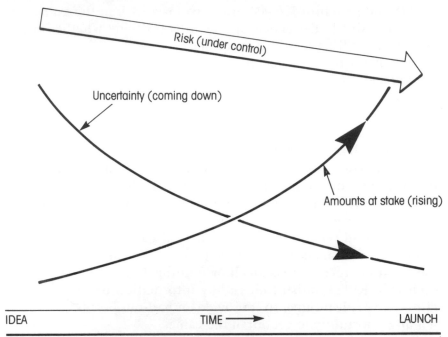

Adapted from R.G. Cooper and B. Little, "Reducing the Risk of Industrial New Product Development," *Canadian Marketer* 7 (Fall 1974): 7–12.

the new-product process — every notch up on the "amounts at stake" curve of Exhibit 3.2 — must bring a corresponding reduction in the uncertainty curve. The entire new-product process, from idea to launch, can be viewed as an uncertainty-reduction process. Remember the five simple rules:

1. When uncertainties relating to the new-product project are high (that is, when prospects of success are fuzzy), keep the amounts at stake low. When you don't know where you're going, take small steps. An all-or-nothing, "damn the torpedoes, full speed ahead" attitude might make a great movie script, but is likely to lead to disaster in the new-product game.
2. As the uncertainties decrease, let the amounts at stake increase. As you learn more about where you're going, take bigger and bigger steps.
3. "Incrementalize" the new-product process into a series of steps or stages; that is, break the "big decision" down into a series of smaller decisions.

Exhibit 3.3.   Risk Out of Control in the New-Product Process

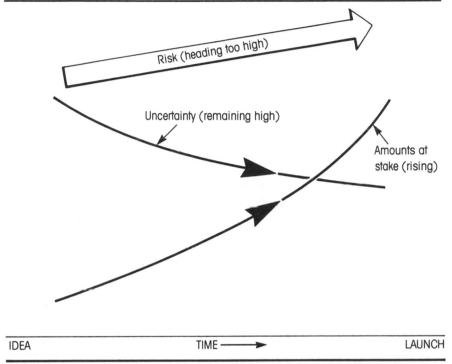

Adapted from R.G. Cooper and B. Little, "Reducing the Risk of Industrial New Product Development," *Canadian Marketer* 7 (Fall 1974): 7–12.

4. View each step as a means of reducing uncertainty. Remember that information is the key to uncertainty reduction. Each step in the process that creates an expenditure must reduce uncertainty by an equivalent amount. "Buy" a series of looks at the project's outcome.
5. Provide for timely evaluation and bail-out points. Ask yourself periodically, "Are we still in the game? Should we proceed or should we kill the project now?" These questions are essential to risk management.

Suggestion: The five decision-making rules outlined above apply to almost any high-risk situation. Does your company follow them in its day-to-day management practices? Review your firm's new-product practices, perhaps using an actual case, and assess whether your management group is handling risk appropriately.

A Seven-Stage Game Plan

These five basic rules of risk management, together with the ten lessons from investigations of past successes and failures, have been fashioned into an effective new-product game plan.

The game plan is a series of moves or plays, and is set up as a stepwise process — a model or scheme to move a new-product project from the idea stage through to market launch.[5] The game plan outlines the many steps and activities that should be built into a successful new-product project: it is a blueprint for the new-product process.

The game plan is based on the experiences, suggestions, and observations of a large number of managers and on my own and others' research in the field. Indeed, my observations of what happened in over 60 actual case histories laid the foundations for the game plan.[6]

The game plan consists of seven main stages, beginning at the idea stage and culminating in the launch stage. The general flow of the process model is shown pictorially in Exhibit 3.4. One further stage can be superimposed over (or on top of) the model: the formulation of a new-product strategy. This strategy formulation stage is omitted from the game plan at present, not because it's unimportant, but because it is "macro" in nature, dealing with the firm's entire new-product program, while the game plan focuses on an individual new-product project. The strategy formulation stage is a prerequisite to the game plan; it is the topic of chapter 9.

Exhibit 3.4.   The Seven-Stage Game Plan

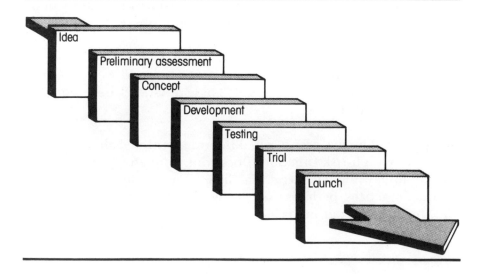

The Structure of the Plan

Each stage of the game plan is made up of a number of specific activities (see Exhibit 3.5). Note that the flow of the plan moves from left to right. The technical and production activities are shown across the top, the market-oriented activities across the bottom, and the evaluation or GO/KILL points across the middle of the exhibit.

Before analyzing the plan on a step-by-step basis, it may be useful to list several of its features:

- The game plan is incremental: the entire new-product process is divided into a series of smaller steps.
- The plan is designed to manage risk: each step is progressively more expensive than the one that precedes it, but the various activities are deliberately designed to drive uncertainty down as one moves from left to right through the model.
- Each stage is separated from the one before it by an evaluation or bail-out point — a GO/KILL decision node that asks the question, "Are we still in the game?"
- The plan is decidedly market-oriented, ensuring that vital and often overlooked market activities are built into the process.
- There is a heavy emphasis on the front-end activities that precede the actual design and development of the product.

Stage 1. The Idea

The game plan begins with the definition of a product idea. An idea occurs when technological possibilities are matched with market needs and an expected market demand. Ideas may be generated by the marketplace — a competitor's new product, recognition of unsatisfied customers' needs, or direct requests from customers. Such market-pull ideas represent the majority of new product projects. But technology-push ideas — which are generated by research or a serendipitous discovery — also play an important role, particularly in radical innovations or breakthrough products.[7]

A good new-product idea can make or break a project. Since it is the product idea that initiates the whole process, there is a strong need for a large number of good new-product ideas. The attrition rate of new product ideas and projects is high, and many firms are bankrupt of good new-product ideas. An important facet of the game plan is the development of an idea-generating system, a topic dealt with in chapter 4.

Exhibit 3.5.   The Detailed Game Plan

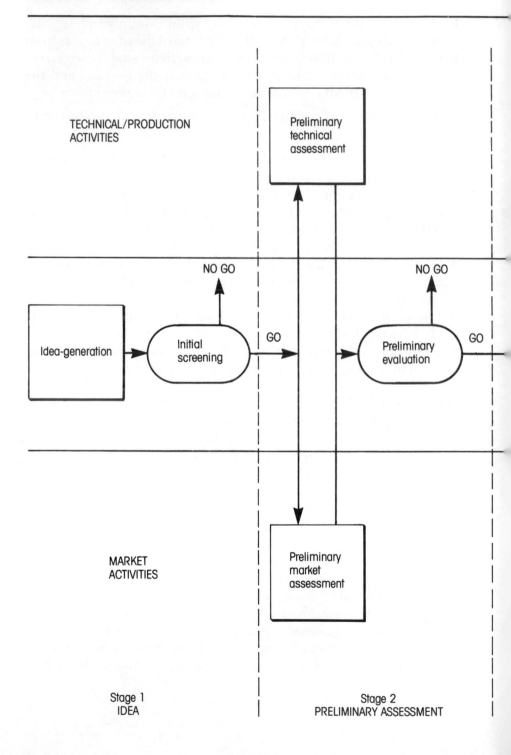

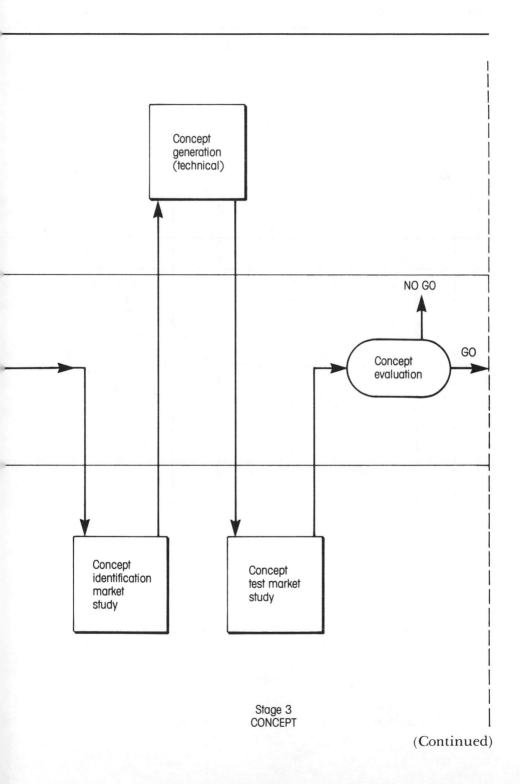

Stage 3
CONCEPT

(Continued)

Exhibit 3.5 (continued)

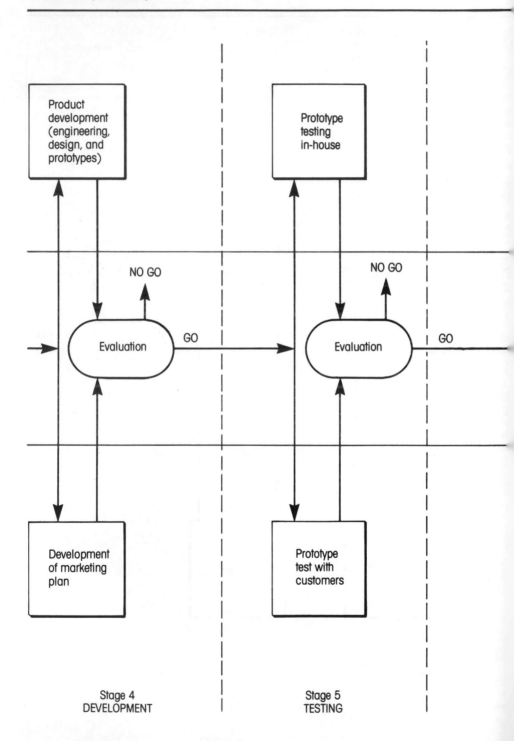

Stage 4
DEVELOPMENT

Stage 5
TESTING

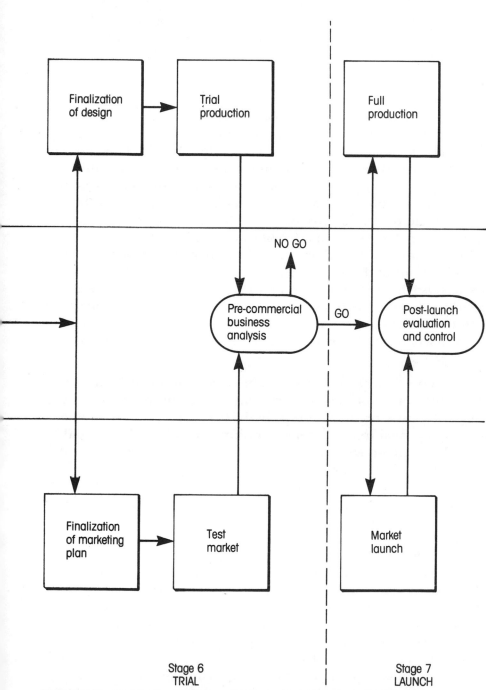

Stage 6
TRIAL

Stage 7
LAUNCH

Suggestion: Review the progress of the new-product ideas in your firm. Where do new-product ideas come from in your company? Where should they be coming from? Are the ideas good ones? How does your company actively solicit new-product ideas? Do you have a new-product idea-generating system in your firm? If the answers to these questions make you uneasy, don't worry; chapter 4 suggests some concrete action that can be taken.

The next part of the idea stage is screening. Screening is the first evaluation of the new-product idea, and represents the first decision to commit resources to the new-product project. Because little information is available at the screening stage, the GO/KILL decision can only be a tentative one. The time and money resources committed to the project at this point are also small. Using this incremental-commitment approach — deciding to spend a little money to "buy a look" at the product's prospects — balances resources and uncertainty, and risk is kept to a tolerable level.

Increasingly, new-product managers recognize that the screening decision is one of the most important steps in new-product process. Yet it is also the least formal, the least consistent, and among the most poorly undertaken. In recent years, however, there has been a move to introduce formal, consistent, and reliable screening techniques in order to separate the good ideas from the bad. This topic is discussed in chapter 5.

What does the screening decision involve? Consider the following illustration. A major chemical company has streamlined the first two stages of its new-product process. In the first stage, product ideas are actively solicited from a number of sources within and outside the firm. These are stored as written proposals. This is called the "white stage." Initial screening is next, and involves the use of a checklist questionnaire completed by people from different areas within the company — R & D, marketing, etc. Responses are combined and weighted by computer to yield a "product score" for each proposal. If an acceptable score is achieved, the proposal becomes a project, and moves to the "green stage" (stage 2 of the game plan).

Consider another example. A multinational firm in the consumer goods business recently reviewed its projects in progress, only to find that too many had slipped into product development without adequate evaluation or rationale. Initial screening was identified as a particularly weak area. The firm implemented a rigorous screening approach whereby all proposals or product ideas must be subjected to a series of "must" questions; if the proposal passes this first cut, it then is subjected to a set of "should" questions. ("Must" and "should" questions are discussed in more detail in chapter 5.) A team of in-house evaluators is briefed on the project; then, independently of each other, the evaluators rate the project on each of 50 criteria. (None of these are financial criteria at this early stage.) Their independent ratings are combined in a weighted fashion to yield a project score and a projected likelihood of success. This is done by comparing the proposal's profile on the 50 criteria against historical successes and failures. The process not only yields a better and more thoughtful early evaluation of the proposal, but is also instrumental in identifying key facets of the project that need more work, time, and information.

Suggestion: Take a close look at how a product idea moves through your company from its conception to the point where time and money are spent on it. Did the idea gradually slip into project status? Who made the decision to implement the idea? How are new-product ideas screened in your firm? Is there a formal and consistent process for evaluating new-product proposals? What are the results? Does your system fail to weed out the obvious losers and misfits? Or is it too rigorous, accepting only the "sure bets"? If you think there's room for improvement in how your firm makes this early selection decision, then chapter 5 will suggest some answers to you.

Stage 2. Preliminary Assessment

Preliminary assessment is the first stage at which significant resources are spent to gather information regarding the feasibility of a project. Expenditures at this preliminary stage, which includes market and technical assessment, should be limited to a specified ceiling. In short, the output of a GO screening decision in Stage 1 can be expressed thus: "On the basis of the information available, this proposal has merit. Spend no more than $10,000 and 15 man-days, and report back to us in one month, armed with much better information, for a more thorough evaluation."

Preliminary market assessment involves a quick-and-dirty market study. The task is to find out as much as you can about market size, growth, segments, and competition by a specified deadline, to determine whether the proposed product has a hope of market acceptance, and to identify and assess sales, promotion, and distribution procedures.

Given its limited budget and short duration, this type of market assessment is clearly not a professional and scientific piece of research. Rather, it's "grunt work": getting hold of available market information in-house (for example, meeting with the sales force); examining secondary sources (for example, reports and articles published by trade associations, research bodies, and government); and canvassing outside sources (such as industry experts or potential customers). It's tough work, much like playing detective and following up on leads, but it's surprising how much information about a new product's market prospects can be gleaned from several days of solid sleuthing.

Preliminary technical assessment involves subjecting the proposed new product to the firm's technical staff — R & D and engineering — for appraisal. Occasionally, outside experts will be used. The key questions concern the technical viability of the product. Can it be developed? At what cost? How long will it take? Can we produce or manufacture it? What resources will be required? What kinds of problems will be encountered?

After the preliminary market and technical assessments are completed, preliminary evaluation can begin. The information provided by the initial assessments will enable you to undertake a more thorough qualitative analysis of the new product. If, when the preliminary evaluation is complete, your decision is to GO, the next and much more expensive stage begins: concept definition.

What factors should be taken into account in the preliminary evaluation? Consider the following illustration. The major chemical firm referred to above has moved the project out of the screening stage into the "green stage," and relatively inexpensive preliminary technical and market assessment is undertaken on the product. With more complete information, a second evaluation is carried out, based on a checklist scoring model and a financial index calculation. If a GO decision is made after this information is evaluated, the project moves to the "blue stage" and more extensive market and technical studies begin.

Consider another illustration. A small firm, Isofab Inc., makes insulation products. The firm wanted to develop a new product, a sound-absorbing brick. The idea came about as the result of some lab work on ceramic sponge-like materials. The product would be used chiefly in highway sound barriers. Management had no idea of market size, pricing structure, or acceptance of a new premium-priced product. Development costs were estimated at $200,000 (a sizeable amount for this small firm). A preliminary market study was initiated on a budget of less than $5,000.

The owner-manager got in touch with a nearby business school, and acquired the services of a graduate student. The student's instructions were to phone every highway department in every state and province in the United States and Canada; to talk to the sound-abatement engineer; and to ask ten minutes' worth of questions. The questions related to the number of miles of past and future installations of sound barriers, types of barriers used, cost per square or linear foot, problems with current products, and the engineer's reactions to the new product idea.

The phone blitz was finished in a week and a half, the student was $1,500 richer, and the owner-manager saved himself a $200,000 disaster. It was discovered that the market was shrinking quickly; current products, although they only reflected and did not absorb sound, were considered adequate (they met federal standards); and the price per foot was of prime concern to the state governments. Clearly, there was little room for a premium-priced product that was perceived as not really necessary.

This example shows that market studies are not the sole domain of the large firm. The costs can be kept down, and one person can gather valuable information from a large number of potential customers or experts in a week or two of hard work.

Stage 3. Concept Definition

Concept definition is the last of the up-front activities to be undertaken before product development begins. It is perhaps the most difficult and certainly the most expensive of the pre-development stages; moreover, it is frequently poorly handled or omitted altogether.

The purpose of this stage is to define exactly the product's concept and the product strategy. This must be done before development work begins. Otherwise, the development group faces a vague product definition, one which is often a moving target. Inherent in this stage is the need to define a winning product concept — to put some meat on the product idea, and to translate the idea into a unique, superior product, with real benefits for the end user.

The objectives of concept definition include the development of a protocol.[8] Agreement must be reached on:

- the definition of the target market;
- the product's concept, the benefits the product will deliver and its positioning in the market;
- the product's attributes and, as far as possible, specifications.

Properly undertaken, this concept-definition stage should clarify the objectives of the product-development team. The result is a more efficient and proficient product design and development effort. But equally important (and often overlooked), an effective concept definition should result in a "winning definition" — in a product that outscores competitors' products, a product with superior benefits for the customer. This winning definition, as we know, is critical to success.

The first activity in the concept-definition stage is usually a market study, which is called concept identification. (Refer to Exhibit 3.5.) This market study of potential users or buyers attempts to identify a niche in the marketplace — a segment of customers who are dissatisfied with what is currently available, a poorly designed or built competitive product, or an area in which a new design or better technology can be used to advantage. The second facet of the study is the clarification of what must be done to achieve success: the identification of the desired benefits or features sought in this new and winning product. Chapter 6 deals with the design of this type of market study; for now, let's look at one example.

A major manufacturer of high-quality highway trucks sought entry into the urban market by developing a dump truck for construction applications. The objective was to build a high-quality, premium-priced truck. A preliminary investigation revealed that the company could indeed design and build such a vehicle, and that the market was a large and attractive one. But the details of the winning product concept and how the product would be positioned in the market were missing. A concept-identification market study was commissioned. Dump-truck fleet owners were personally interviewed to determine their choice criteria, their preferences among various makes of trucks and the reasons for those preferences, and the design improvements that could be made. Following the interviews, it became clear that truck downtime was a major concern and a costly problem to fleet operators. In fact, some users kept spare vehicles simply to ensure that they could meet their customers' demands. Interviews also revealed that most maintenance was done at night and that feverish efforts were made to get the vehicles ready for work the next morning. So the target market was defined in terms of a "benefit segment" — those construction fleet owners for whom downtime was a critical and costly problem. The winning product concept was defined as "a truck that can be repaired and back on the road in 12 hours no matter how serious the breakdown."

The design requirements for what constitutes a better product for the customer are defined from the concept-identification market study described above. Next comes concept development (see Exhibit 3.5). There the market requirements are translated into an operational concept, one that is technically feasible. The truck manufacturer's engineers, armed with the desired concept of the "minimum downtime truck" or the "12-hour truck," conceived a modular vehicle. Every major "trouble component" in the truck was engineered to be removed and replaced quickly and easily — engine, transmission, rear axle, front axle, electrics, radiator. Standard parts available from a variety of local suppliers were used. The "12-hour" concept was found to be technically and financially feasible.

The final activity in the concept-identification stage is a concept test — a test of the likely acceptance of the product by the marketplace. This market study of potential buyers, different from the "prospecting" concept-identification study, presents the technically feasible concept to customers and measures their reactions to it. Usually this presentation is made in the form of sketches, diagrams, models, or written descriptions of the product. (Because the product has not yet been developed, no prototype or sample is available.) The object of the study is to gauge market acceptance of the new product: the customer's interest, liking, preference, and intent to purchase. Additional information gleaned from such a study may involve suggested modifications to the proposed concept, or, in the event of rejection, the reasons that the concept is unacceptable to the buyer.

For example, before proceeding with commercial development, a manufacturer of health-care products conducted a concept test of a total health-testing unit aimed at executives in corporations. A van equipped with advanced equipment could be used by health-care professionals to perform a complete medical examination in minutes. The van was to be brought directly to the corporate premises and executives examined on-site. The benefits to the corporate user were thought to be better executive health and the elimination of the need for inconvenient two-day annual checkups in a hospital. For the market-concept test, artist's sketches, simulated promotional material, and a miniature model of the van's interior were used. The visuals, descriptions, and price information were displayed to decision-makers in potential client firms, and their reactions — interest, liking, and purchase intent — were gauged.

The two market studies at this concept stage do more than identify and test a product concept. They also begin the marketing-planning process. These up-front activities define the target market, specify the core element of the marketing mix — the product strategy — and result in a product-positioning statement.

Although the design and market acceptance of the product is usually the principal concern of the market studies in stage 3, questions about how the product will be sold, promoted, and distributed can also be addressed. It takes little extra effort to add a section to the questionnaire that asks questions about buyers' purchasing habits. Information gleaned from such questions is vital to the design of a launch plan for the product.

A concept-evaluation decision is now made (see Exhibit 3.5). Note that the concept test provides intent-to-purchase data which permit estimates of market acceptance and expected sales. Similarly, the concept development provides estimates of development and production costs. For the first time in the project, a reasonable financial analysis can be performed. And a reliable evaluation is essential at this point, before development work begins.

Suggestion: Take a close look at those activities that occur just before the product-development stage in your firm. (For example, review some actual new-product projects.) Have you built in the necessary market-research investigations that probe the customer's needs, wants, and preferences in order to determine what the winning new product concept should be — the benefits the product should deliver to the user, how the product should be positioned, and what features and attributes should be in the product's design? What steps do you take to ensure that your proposed product definition is indeed acceptable to the customer? Do you regularly undertake concept tests in which potential customers are exposed to the proposed new product's design before development work actually gets underway? By the time you enter the development stage, how well have you defined the product? Is the R & D team expected to arrive at a product design in a relative vacuum? If some of these pre-development activities are not a regular facet of your new-product process, chapter 6 offers some guidance in implementing them.

Stage 4. Development

In stage 4, the actual product development begins in earnest. Technical resources people — the design, R & D, and engineering departments — are deeply involved, and a prototype or product sample is the usual outcome of this stage.

Paralleling the development of the product is the development of the marketing plan. (Refer to Exhibit 3.5.) Here the results of the concept stage — target market selection, product strategy, and positioning strategy — are shaped into a marketing plan. Next, the supporting elements of the marketing mix — pricing, distribution, advertising, sales strategy, promotion, and service — are decided. These supporting elements may require an additional market study of how customers buy the product, what influences their purchases, and the customer's source of product information. Chapter 8 focuses on the design of the marketing plan.

Stage 5. Testing

The testing stage is a validation of the product under both simulated and real-world conditions. In-house prototype testing involves subjecting the product to a variety of tests under controlled conditions to ensure that no technical flaws exist.

At the same time that the prototype testing is undertaken, a customer test of the product is conducted. The objective is to test the product under actual use conditions to ensure that the product performs as expected. Customers have an innate ability to find new ways to destroy products, ways that even the most thorough lab tests could not anticipate. A second purpose of the customer tests is to gain a more realistic appraisal of the product from the customer's standpoint; customer tests can be used to gauge liking and preference. The outcome of these tests is the identification of defects and needed improvements, and expected customer acceptance of the new product.

For example, during the development of a dial-in-hand telephone for customer use, a major telephone manufacturer assembled 100 prototype units.[9] Fifty of these were used for in-house tests of reliability and durability. The other 50 were installed in customers' homes. The customer test proved crucial to the product's eventual success. A potentially disastrous flaw in the design was uncovered. In the wall-phone design, the receiver fell off the hook when a nearby door was slammed hard enough to jiggle the wall in a stud-wall-construction house. The lab, of course, had concrete block walls, and the problem went undetected until the customer test. A minor design modification to the receiver overcame the problem before thousands of phones with faulty receivers found their way to households.

After the completion of in-house and customer testing to verify the product's design and its market acceptance, another evaluation is made to gain further and better estimates of production costs, marketing costs, and customer acceptance. A realistic financial projection and appraisal can now be made. A GO decision moves the product to stage 4, the trial stage.

Suggestion: Is a customer test (or a preference test involving the customer) a regular feature of your firm's new-product process? For most firms it is, especially in the consumer-goods industry. But industrial-goods manufacturers can also benefit from more customer testing. If you want proof, review the last few new products you launched that faced serious difficulties in the marketplace. Had you built a customer test into the project, including measurement of customer reactions, liking, and intent to purchase, how many of these problem products could you have spotted well before they were into the launch phase?

Stage 6. Trial

The trial stage represents a dry run of all commercial facets of the project: production, product design, and marketing. It is the first point in the project where everything comes together. Before the trials begin, however, both the product design and the marketing plan must be fine-tuned and completed. The results of the customer tests and the in-house tests provide the information necessary to finalize the product design. The development of the marketing plan has been proceeding since the concept stage, and it too should be close to finalization by now.

A pilot production run tests the production method that will eventually be used for full-scale production. Modifications to the final production facilities or methods are often required in order to solve unexpected problems.

A test market — selling the product using the proposed marketing plan, to a limited number of customers or in a limited geographic area — tests not only the product but also all the elements of the marketing mix. A pre-test market, which is a lower-cost simulation of a test market, can also be used for some categories of new products. The identification of needed adjustments to the marketing plan and final estimates of market share and sales volume are two results of the test market.

The telephone manufacturer in our earlier example used a trial production run of 1,000 new dial-in-hand phones to obtain true estimates of production costs and to spot and correct production problems. The 1,000 phones were then sold in a test market with the cooperation of the manufacturer's own customer, the local telephone company. All the elements of the marketing mix were tested, including the pricing strategy (an additional monthly premium on the phone bill); the communications strategy (newspaper, radio, and announcements enclosed with the phone bill); and personal selling (the use of installers to promote the product in new installations). The test market confirmed the efficacy of most of the elements, but revealed that sales would be double what the manufacturer had thought. A revised national launch plan involving introducing the product on a region-by-region basis rather than nationally was quickly devised, so that demand would not far exceed production capacity.

Before entering the launch stage, a final pre-commercialization business analysis and evaluation are made based on the concrete financial data derived from the test market and trial production. The product is now GO for launch.

Stage 7. Launch

The launch stage involves the start-up of full or commercial production and the implementation of the marketing plan in the total market area. All facets of the launch plan have been well tested. If the tests have been properly carried out, and barring any unforeseen or new circumstances, the launch should be a simple matter of proficiently executing a well-designed plan of action.

Post-launch evaluations that take place at designated times provide benchmarks to gauge whether the product is on target. These benchmarks may include assessment of actual and projected market share, sales volume, and production costs. These criteria will have been established before the launch. The evaluations — where the product is in the market versus where it was intended to be — are essential to post-launch management, and can signal the need for corrective action to move the product back on course.

The Game Plan: A Recap

New-product development and launch will always be a high-risk undertaking. Nonetheless, much can be learned about effective new-product management from a review of past projects and the experience of other firms. Many of the insights and rules relating to effective risk management have been incorporated into our seven-stage game plan. No new-product project will necessarily follow the model religiously; unforeseen events and special circumstances often will dictate additional steps or repetition or omission of other steps. But this game plan does provide an outline of the process — an outline that ensures that the critical steps are not overlooked. Before you deviate too far from the ideal game plan, make sure that you are aware of the basic steps.

The Benefits of the Game Plan

The benefits of implementing the game plan are many. One valuable result is that the new-product process becomes more multidisciplinary. The technical and production activities are shown at the top of the model in Exhibit 3.5, and the market-oriented activities are shown at the bottom. The balance between the internal and the external orientation is obvious. If all the steps in the game plan are carried out, the process demands multidisciplinary inputs. The simple fact is that no one player can carry out all the varied steps. Many players with different skills must be involved. Notice, for example, the amount of crisscrossing back and forth between marketing and technical activities throughout the model. The process forces cooperation and exchange between marketing, R & D, engineering, design, production, and finance at virtually every stage in the project — clearly an essential element in a successful new-product program.

A second benefit of the game plan is its risk-management nature. Earlier in this chapter five rules of risk management were highlighted — incremental commitment, balancing expenditures with uncertainties, buying information to reduce uncertainty, and timely evaluation and bail-out points. The game plan incorporates all of these rules. It is incremental; decisions to commit are made a step at a time. It is designed to reduce uncertainties as expenditures increase. Information, particularly market information, plays a key role in the game plan, and is used to reduce uncertainties at every stage. And evaluation and bail-out points — the GO/KILL decision points — force constant attention to the question, "Should we still be in the game?"

A third benefit of the game plan is its incorporation of many of the elements essential to success. There is a strong commitment to developing a unique, superior product that delivers real benefits to the user in the concept-development stage. A market orientation is prevalent throughout the game plan and at every stage of the process, not just at the launch phase of the project. The up-front or pre-development activities — idea, screening, preliminary assessment, market studies, and concept definition — that are lacking in many firms play a dominant role in the game plan. Project evaluation, particularly at the early stages, is also stressed. The marketing-planning process begins early in the process to ensure a well-conceived launch of the product into the market.

Finally, the game plan is based on a recognition that most new-product projects fail; that management is usually to blame; that success depends on doing many things right; and that a single miscue can spell disaster. The game plan brings discipline into what has traditionally been an ad hoc management area.

No Guarantees

New-product success can never be guaranteed, but thoughtful attention and a systematic approach to the conception, development, and launch of new products can help us avoid many of the pitfalls that have plagued our product development efforts in the past. Don't ask whether you can afford to implement the game plan. That's the wrong question; the right question is, "Can you afford not to?"

Finding Good
New-Product Ideas

The Challenge

What's a good idea worth? Everything, if you believe some new-product managers. That may be an overstatement; mediocre ideas can often be built on to yield superb new products, and there are countless examples of good ideas that came to nothing. But one point needs making: the new-product idea initiates the process and the idea shapes much of the rest of the new-product effort. If management has only ho-hum ideas to choose from, and few of those, don't expect miracles in the new-product program.

The generation of new-product ideas is the first step of the new-product game plan. A failure at this stage places the rest of the process in jeopardy. One problem is that there is a very high attrition rate of product ideas. In chapter 2 we saw that most new-product ideas never make it to market; for every seven new-product ideas that enter the process, only one becomes a commercial success. So quantity as well as quality of ideas is important.

Several years ago I served as a consultant to the corporate new-products group in a multi-billion-dollar consumer-goods firm. One day was to be spent discussing the new-product game plan, with an emphasis on project selection. Partway through the discussion, I was met with blank stares and uninterested looks. I finally asked, "How do you screen new products now, and what's the problem?" After a moment of silence, I was told, "We don't. We take any idea that comes along. We simply don't have enough ideas to keep us busy." How many other companies are bankrupt of good new-product ideas?

The challenge is to find new-product ideas — good ideas, and lots of them. The problem is that good ideas don't grow on trees. Traditionally, firms and individuals have expected new product ideas suddenly to appear, almost magically, from thin air. Occasionally, one gets lucky and ideas do come out of nowhere. But if the objective is to have continuity — many good ideas year after year — then relying on Lady Luck as an idea-generator is foolhardy. More and more managers recognize that ideas don't just appear, certainly not in the quantity and quality desired by the firm. But they can be made to happen! What is needed is a conscious, deliberate, ongoing system of generating new-product ideas.

Suggestion: Let's take a practical example. How many new-product ideas did your firm or division screen last year? Write the number down. For a medium-sized firm, the number might be one hundred. Now, here's the challenge: you want four times that number screened by next year. How to do it? Try writing down three specific things you or your colleagues can do to quadruple the number of ideas. Next question: how many of these idea-generating activities are you performing today? The sad part is that most of us can think of a number of specific schemes, activities, and techniques to stimulate the generation of new product ideas. Most cost little money and take little effort to implement.

This chapter will offer guidelines for developing an idea-generating system that will increase the number of new product ideas by a factor of two, three, or four. A more appropriate title for this chapter might be "Twenty Ways I Can Triple My Number of New-Product Ideas." (If you are already blessed with more good-quality ideas that you can handle, you're one of the lucky few. Skip this chapter and move on to chapter 5, where you'll find out how to screen the best ideas.)

Establishing a Focal Point

A problem in many firms is that no one is changed with the task of idea-generation. It's everyone's job and no one's responsibility, so it rarely gets done in an organized and consistent fashion. An "idea person" is needed in the division or company.

Even when ideas do surface, often the proponent doesn't know what to do with the idea — there's no one to send it to for action. There is no focal point, no idea person, in the firm. The situation was so bad in one firm that management installed a dedicated phone line with a taped message: "At the sound of the beep, please state your new product idea." A little bit impersonal perhaps, but a recognition that this company lacked a "receiving station" for new-product ideas.

What of the sales force? Picture a young salesman, relatively fresh from university, still creative, working in a sales office thousands of miles from headquarters. In the course of a year, he will almost certainly come up with one new-product idea. But what does he do with it? Whom does he phone? If no one has been identified as the receiver of ideas, that salesman's idea will be lost forever. Multiply his case by the number of sales representatives in your firm, and you'll see the magnitude of the problem.

The first step in setting up an idea-generating system is to assign one person the responsibility of stimulating and generating new product ideas. He or she can be located in one of many groups in your firm. If you have a new-products department, that is the logical location for the idea person. Failing that, the person can be in marketing, engineering, or R & D. The exact location is not important; what is important is defining the job and assigning the right person to it.

The idea person should be given an objective of a specific number of ideas for next year. He or she is charged with determining how to achieve that objective. The existing and potential sources of ideas within and outside the firm must be identified, and lines of communication and methods of idea-generation, or "flow lines," must be established. (Suggestions for facilitating this flow are given later in this chapter.) Of course, the idea person logically becomes the focal point for ideas — the person one phones or writes to about a new-product idea. This is important, for without a focal point ideas are likely to get lost. Having a person in place — a face, a name, a phone number, a personality, someone whom people in the organization know — is a vital step in implementing an idea-generation system.

Implicit in this role of focal point is action; the idea person moves the idea to the next step of the process, idea screening and evaluation. He or she must also be able to explain to employees or clients what a viable new-product idea is. If the characteristics of a legitimate new-product idea are not defined beforehand, expect a mixed bag of ideas: the quantity might be there, but quality will suffer. These requirements should come from the corporate new-product strategy (the topic of chapter 9), and from screening criteria (the topic of chapter 5).

Suggestion: Appoint one person charged with stimulating and facilitating the flow of ideas from a variety of existing and potential sources. Set objectives for the idea person. He or she will become the focal point for new-product ideas; make sure that your potential sources of ideas know how to reach him or her.

Identify the Sources of Ideas

Where do new-product ideas come from in your company? Where should they be coming from? Exhibit 4.1 illustrates major sources of new-product ideas. Not surprisingly, the main sources are the sales and marketing departments and the R & D and engineering departments. More than half the ideas originate in these areas, with about equal frequency. Less obvious sources may also contribute a significant number of ideas, however. Customers are an excellent source of new-product ideas, and in some industries they take an active part in product development. According to Von Hippel, the majority of new products in certain industries, such as scientific instrumentation and electronic equipment, are not only suggested but actually developed by the customer.[1] Exhibit 4.2 shows the breakdown between user-developed and manufacturer-developed new products in selected industries. Von Hippel concludes that when the largest benefits accrue to the user (as opposed to the manufacturer), the user will be the major source of ideas and new products.

Suggestion: Identify the existing and potential sources of ideas in your firm. Make a complete list of these sources.

Exhibit 4.1.   Primary Sources of R & D Ideas in 40 Companies

INTERNAL SOURCES	NUMBER OF COMPANIES
Research and engineering	33
Sales, marketing, and planning	30
Production	12
Other executives and board of directors	10

EXTERNAL SOURCES	
Customers and prospects	16
Contract research organizations and consultants	7
Technical publications	4
Competitors	4
Universities	3
Inventors	3
Unsolicited sources	3
Advertising agency	2
Suppliers	2
Government agencies	2

Adapted from B.V. Dean, *Evaluating, Selecting, and Controlling R & D Projects*, research study no. 89 (New York: American Management Association, 1968).

Exhibit 4.2. Innovation Sources: User Versus Manufacturer

INDUSTRY	INNOVATIVE PRODUCTS FIRST DEVELOPED BY	
	USERS	MANUFACTURERS
Semiconductor and electronic subassembly process equipment	67%	33%
First of type (innovation)	100%	0%
Major functional improvements	82%	18%
Minor functional improvements	70%	30%
Scientific instruments	77%	23%
First of type (innovation)	100%	0%
Major functional improvements	63%*	21%*
Minor functional improvements	20%*	29%*

* The missing percentages represent joint user-manufacturer innovations.

Source: E.A. Von Hippel, "Has Your Customer Already Developed Your Next Product?" *Sloan Management Review* (Winter 1977): 63–74. See also J. Kreilling and E.A. Von Hippel, "Users Dominate Much Instrument Innovation," *Instrument Technology* (Feb. 1979): 7.

Establishing Flow Lines

When the sources of ideas have been identified, the next step is to develop and implement ways of stimulating and facilitating the flow of ideas from these sources to the focal point. Let us begin with external sources.

Customers

Customers represent a huge and often untapped potential source of new-product ideas. Most successful new products are market-derived. Other than relying on the sales force to pass suggestions on to the new-product idea person, what approaches can be used to tap this potential? Some suggestions follow:

1. Use "focus" groups of customers. Try mounting several group discussions (professionally moderated) with groups of invited customers or potential users. The group discussion focuses on problems customers have with products in your product category. This problem-oriented discussion can then turn to pinpointing possible solutions, resulting in valuable customer-derived ideas.

There are a variety of other group techniques, such as brainstorming, synectics, etc. — commonly called "creativity methods" — that have applicability not only to customer groups but also to in-house groups. These methods are discussed later in this chapter.

Focus groups of customers are most frequently used by consumer-goods firms to screen new product ideas. Note, however, that the method can and should also be used to advantage by industrial-goods producers as a means of idea-generation. One major electrical-equipment manufacturer arranged a weekend for a management retreat. The purpose was to identify new business and new-product opportunities. Almost as an afterthought, half a dozen key industrial customers were invited. The customers enthusiastically accepted the invitation, and in fact proved to be the source of the most original and promising ideas.

2. Set up customer panels to generate ideas. A permanent panel of selected customers that meets on a regular basis is another good source of ideas. One major manufacturer of sporting-goods equipment uses customer panels successfully. After admitting that his firm was short on new-product ideas (how many new ideas in baseball gloves, hockey pads, golf bags, and soccer balls can you think of?), the company president organized panels made up of coaches of minor-league teams. He recognized that most team-sport innovations took place in minor and little leagues, often for safety reasons. By listening to the suggestions of the coaches, he was able to identify problems and possible solutions leading to new-product ideas.

In the packaged-goods business, Procter & Gamble uses panels of customers who are regularly questioned about possible product improvements and modifications. This type of customer-panel arrangement can be used as successfully by small firms as by multinationals.

3. Survey your customers. Pick a random sample of customers, develop a research questionnaire, and visit or phone the customers to determine their needs, problems, and desired product solutions. One manufacturer of fiberglass materials undertook such a survey prior to embarking on a major product-redevelopment program. Surprisingly, the market study was commissioned by the R & D department, not by marketing. Research and development personnel visited about 60 customers, mostly fiberglass fabricators, and spent a day at each factory observing how the customers used the company's and competitive products. They questioned production workers and conducted an in-depth interview with management using a lengthy questionnaire. The result was a much sounder knowledge of customer problems and the identification of a number of ideas and features to be built into the new line of products.

4. Observe your customers. Observation is the most underutilized of all market intelligence techniques. Ideas for new or improved products come from watching how customers use current products and from ascertaining the problems they have with them. One major consumer-products firm relied heavily on observation to generate new-product ideas. An independent "researcher" visited homes and watched homemakers doing their daily chores. These visits and the discussions that ensued resulted in a number of worthwhile product ideas. The technique is no longer used by the firm; the researcher retired, and was never replaced. But experienced marketing personnel at this company reminisce about this unique, albeit unsophisticated, means of idea-generation.

More than one farm-implement manufacturer has uncovered brilliant new-product ideas simply by watching farmers going about their daily farming activities. Trudging through the barn or field in hip boots is not very glamorous work for the researcher, but nobody ever said the best ideas would be generated in a fancy office on the fifteenth floor at headquarters. Observation works — try it.

5. Install a customer hot line. Put in a toll-free number so that customers can call in with their suggested improvements. Advertise the existence of this hot line on your package or product literature. One large consumer goods company discovered this approach by accident. The toll-free number was installed to handle customer complaints; it turned out to be a remarkably good source of new-product ideas.

6. Other approaches. Von Hippel's work on the customer as a source of ideas and new products leads to a number of specific suggestions.[2] Von Hippel sees customers not only as a source of new-product ideas, but as a source of partially and sometimes completely developed new products. First, two key characteristics of user-developers must be recognized: a large user population develops relatively few products, of which only a small number will be commercially promising; and user-innovators have no incentive to take their innovations or ideas beyond their own doors. Von Hippel's suggestions include the following:

- Get into the market with a standard product of interest to users who are also innovators — anything that will allow you to establish a sales-and-service relationship with the right group of user engineers.
- Hire people who can recognize potential new products in addition to selling the standard line. Typically, Von Hippel found that

most manufacturers' sales and service people were not trained or equipped to do this. The few who were took the initiative and asked users, "What have you done lately that's new and useful?" Usually the users were happy to explain or demonstrate their innovations.

- Organize your new-product development group so that there is a free flow of information between the sales and service departments.
- In dealing with user-innovators, define the desired product as precisely as possible. For example, companies seeking user-developed software frequently complain that the software they are offered is poorly documented. But the same companies often fail to specify the level and type of documentation they require.
- Specify an appropriate "reward" for users. Don't be a piker! If it will cost you $10 million to design a computer in-house, you can well afford to offer $5 million for a suitable and complete design from users. No matter what your product, offer your reward to users in an appropriate form. Cash prizes are often most desirable to consumers. For example, Pillsbury offers cash awards to the winners of its annual Bake-Off. For individuals in corporations, a cash award may be unacceptable, but there are other types of rewards. For example, Technicon offers a grant to university researchers for commercially promising instruments.
- Target likely innovators only. Aiming rewards at the most probable user-innovators saves advertising expense and reduces in-house screening costs by reducing the number of inappropriate responses. Likely targets might include your own customers, your competitors' customers, or experts in a given area.

Suggestion: A number of concrete steps you can take to get new-product ideas from customers are outlined above. Skip back a few pages and read the list of suggestions again. Ask yourself how many of these steps your firm regularly takes? How difficult will it be to implement the ones that you're not using now?

Competitors

Competitors represent another valuable source of new-product ideas. The objective is not to copy your competitors — copycat products have a much lower chance of success — but to gain ideas for new and improved products from competitors. Often, the knowledge of a competitive product will stimulate your team in arriving at an even better product idea.

Routinely survey your competition. Periodically perform a complete review of competitive products, particularly new ones. Obtain a sample of your competitor's new product. Sometimes a one-for-one trade will work (the competitor is interested in your products too). If not, a trusted customer can be used to obtain a sample of your competitor's product. If all else fails, toss the problem to your top salesperson: he or she is probably quite expert in getting hold of competitors' products.

Once the sample is obtained, undertake a thorough evaluation of the product from a technical standpoint. Arrange an internal brainstorming session aimed at improving on your competitor's product. Determine how well the product is doing in the marketplace from published data, from an industry association, or from your salespeople and customers. Finally, obtain copies of the advertising and promotion material for the product; knowing what the competitor is emphasizing or how he is positioning the product can yield new insights for your own products.

If you're in the service business, a good way to obtain a sample of your competitor's new service is through "mystery shopping." Have one of your staff pose as a potential customer of the competitor, take notes on the features and attributes of the service, and gather any pertinent literature. A word of caution, however: although many firms use mystery shopping, an ethical question arises. How far do you go in the charade? A good rule of thumb is to get as much information as you can, but don't make the competitor's salesperson do a lot of work. If he gets suspicious, be honest; tell him that you're simply doing market research.

Trade Shows

Trade shows are generally viewed as a marketing tool, an opportunity to display products to prospective customers. From a new-products standpoint, the value of a trade show is its potential for generating new-product ideas. Trade shows present the perfect opportunity to solicit dozens of ideas at relatively little expense. Where else can you find all that's new in your field displayed for public consumption? And where else can you find customers by the dozen ready to give you their opinions on new products presented at the show?

Organize a trade show visitation program. Get a list of the relevant trade shows in your industry. Arrange to have at least one person visit each show, even if your firm is not displaying, for the sole purpose of getting new-product ideas. This should not be a social event, but a serious intelligence mission. Arm your intelligence officer with a sketch pad, a notebook, a list of key exhibitors, and perhaps a small camera. His or her task will be to visit each of the key exhibit booths and to itemize and describe the new products on display there. Sketches, notes, brochures, and photos add detail to this description. Your intelligence officer's task is to make a formal presentation to the rest of the new-products group — to present an illustrated review of new products on display at the show. If obtaining brochures is a problem (many shows now use plastic name cards and an addressing machine for sending literature later), consider using a mystery shopper.

Not too long ago, I attended a two-day corporate planning meeting of a large industrial firm. One divisional manager after another stood up and presented the bad news: our market is stagnant, the competition is fierce, our product line is mature, and there are few new opportunities. But the manager of the plastics division (plastic piping and electrical conduits, also a relatively mature area) outlined one new product after another that his division had successfully introduced in the last year, and several more than were under development.

During the coffee break, I chatted with him. "Where do you come up with all these neat new products?" I asked. His reply was simple: "From the Chicago hardware show. Every year my sales manager and I spend two to three days pounding the pavement at that show. We go from booth to booth, just looking for ideas. And every year we come back with a fistful of new-product opportunities. Many of the exhibitors are small firms, often with only local distribution. Others are foreign firms, and lack a marketing effort here. So the market's still wide open for these products." He then went on to describe how several of his division's most successful ideas had originated at that show, and how minor improvements and redesign by his staff had led to the development of unique products.

Trade Publications

As most intelligence officers will attest, the majority of "intelligence information" is in the public domain — it's just a matter of gaining access to it in a regular and coordinated fashion.

Trade publications report new-product introductions. Most manufacturers announce their new products to the trade in these publications, either via advertisements or in editorial copy. Like a trade show, these publications can provide the stimulus for your group to conceive an even better idea. Foreign trade publications show products that may take years to reach your own shores and market, and these too represent opportunities.

The problem is that no one person has the time to read all the relevant trade publications, both domestic and foreign, each month. And even if the publications are read and ideas spotted, what happens to the ideas? Where do they go for action? There are two possible solutions, and both are simple:

- Split the reading task among a number of people. Designate each person to be responsible for reviewing certain trade publications. He or she can prepare a list of products or ideas every month for distribution to the rest of the new-products group. The list, with brief descriptions of the product idea is also sent to the "idea person" for possible action.
- Set up an internal clipping service. Appoint a clerical person to review the magazines and clip new-product announcements. The copies of these clippings can be circulated to the new-products group and to the idea person. In some cities, outside clipping services are commercially available.

Patents

The files in the U.S. patent office contain millions of American and foreign patents, and thousands more are added annually. To keep abreast of developments, and to stimulate new-product ideas, some firms keep a close eye on the weekly Official Gazette, which provides a condensed description of patents issued.

Some years ago, a highly successful entrepreneur in a Canadian firm conceded that most of his successful new-product ideas were the result of a regular review of patents issued in his industry. His firm manufactured chemical and pharmaceutical intermediates, a very competitive and dynamic business. His regular review of issued patents, coupled with a review of trade statistics obtained from government sources, were at the heart of his idea-generation and screening system. Taking advantage of the fact that Canadian patent law differs from U.S. law, he set about making the same product, but via a different chemical process. An unorthodox procedure, perhaps, but for this small businessman regular reviews of issued patents were the key source of this new products and ultimately of his company's success. A few years later, he sold his business to a multinational in the food and beverage field for a handsome price. One wonders if they've discovered his secret.

Brokers

The trade in ideas is big business. Over the last several decades an army of middlemen has sprung up — people in the idea trading, selling, and licensing business. The number of "go-between" firms precludes a complete listing here; many are local organizations or specific to one industry. How many of these idea brokers is your firm in regular contact with? If the answer is none or few, find out who the appropriate brokers are in your field, and make contact; get on their mailing lists and assign someone the task of regularly reviewing their offerings. There are several types of middlemen:

- Patent brokers: firms or individuals specializing in marketing individual inventors' products to manufacturing and service companies.
- License brokers: firms that market licenses to manufacture and sell products. In some cases, the firm has developed a product, but has no real interest in manufacturing and selling the product itself. In other cases the firm is a foreign one, and simply lacks the interest or resources to market in a trading area. Both are sources of good new-product opportunities, and often the ideas and arrangements are handled via middlemen. Some of the larger broker firms publish newsletters listing products available for licensing.
- Licensing shows: some industries organize trade shows whose sole purpose is to present new-product opportunities available for license to prospective manufacturers. Such shows are not only a source of fully developed products, but are also a way to make contact with various middlemen. Of course, these shows are also a stimulus to idea-generation — simply seeing others' new products will often enhance your own creative abilities.

Experience with idea and license brokers is mixed at best. Some managers in small-to-medium-sized firms swear by them. Such firms, which typically operate only locally or nationally, can gain access to new products and ideas, particularly from other countries. But remember that idea middlemen are in business, too — the business of promoting ideas and licenses on behalf of their clients. Some of the ideas they offer may be overstated, and many will be of little or no interest to you. While this source of ideas may not be the most fruitful for every firm, it merits investigation.

Suppliers

Suppliers are often a good source of new-product ideas and help. This is so particularly when the supplier is a large firm with well-funded R & D and applications-engineering facilities. Suppliers too are looking for new applications for their products, and often come up with ideas for their customers.

For example, plastic and fiberglass suppliers provided General Motors with considerable help in the design of the Pontiac Fiero, which involved plastic and fiberglass panels attached to a metal space frame. Similarly, in the heavy-duty truck business, one supplier of cabs and cab components has actually designed a state-of-the-art truck cab, far in advance of anything that Ford, Mack, Kenworth, or Navistar has, in the hope that part or all of the cab would be adopted and built into one of the major manufacturer's own products.

If your firm purchases such materials as steel, glass, fiberglass, metals, paper, or large components, chances are that one of your suppliers is working on new products and concepts. Have your technical and marketing people regularly visit his lab and engineering facilities; establish a liaison with the supplier's own technical people; and, if possible, get on the supplier's mailing list so you'll be kept abreast of developments.

Private Inventors

Yet another source of new product ideas, and sometimes of fully developed products, is the private inventor. For example, Black and Decker makes use of home handymen, who often have new-product ideas and inventions worthy of investigation. The problem with this source is the legal risk — how to handle the unsolicited idea in the context of the law relating to patents and trade secrets. For example, if your firm is working on a product and receives an unsolicited idea for the same product, you may stand to be accused of stealing the idea from the small inventor. At worst, you will face a legal fight; at best, a public-relations embarrassment. As a result, some firms simply refuse to deal with private inventors or unsolicited ideas. This is unfortunate, because it can result in missed opportunities. And, in some circumstances, even a refusal can land you in legal difficulties. For example, if you fail to acknowledge the unsolicited idea, or if you read the unsolicited proposal, you may be accused of having benefited from the idea.

Exhibit 4.3.   How Firms Handle Unsolicited Submissions

	FOODS	HOUSE-HOLD	CLOTHING	PERSONAL	MISC.	TOTAL
LEGALLY SOUND						
Waiver	7.0%	22.2%	21.4%	16.7%	22.5%	17.5%
Patents only	4.7	0	0	14.1	2.5	2.4
Reject outside	27.9	0	0	25.0	7.5	12.6
Total	39.6	22.2	21.4	55.8	32.5	32.5
DANGEROUS						
No response	25.6	42.2	35.7	25.0	15.0	28.3
Evaluated	34.9	35.6	42.9	29.2	52.5	39.2
No good	9.3	17.9	28.7	4.2	10.0	12.7
On market	9.3	4.4	7.1	12.5	25.0	12.0
Known to us	14.0	4.4	0	8.4	10.0	8.5
Not our area	2.3	8.9	7.1	4.1	7.5	6.0
	60.5%	77.8%	78.6%	54.2%	67.5%	67.5%
Number of firms	43	45	14	24	40	166

Foods: Cookie, sandwich, candy, vegetable, cracker, coffee, wine, soft drink, pet food.
Household: Tool, appliance, carpet, television, air freshener, oven tray, cooking bag, bug killer, sound system.
Clothing: Shirt, dress, shoes.
Personal: Hair care, after-shave, shampoo, jewelry, tobacco.
Miscellaneous: Photography, sports, banking, game, auto part, motor additive, office tape dispenser, pen.

Reprinted with permission from C.M. Crawford, "Unsolicited New Product Ideas: Handle with Care," *Research Management* (Jan. 1975): 22.

Crawford studied what happened when 35 unsolicited ideas were submitted to 166 companies.[3] Exhibit 4.3 shows his findings. About one-third of the ideas were handled in a fashion that legal authorities felt was legally sound, and two-thirds were handled in a legally dangerous way. Crawford recommends the following procedure for handling unsolicited ideas.

- Get the submission in writing. Refuse to accept oral submissions.
- The person who receives the submission should forward it immediately to the firm's legal department. As soon as the recipient realizes that the document contains an unsolicited idea, he should stop reading it and turn it over to the lawyers.

- The legal department should give the submission a number or identifying code. A form letter should be sent to the submitter expressing thanks, indicating the conditions under which the firm considers unsolicited submissions, and that a waiver form must be signed and returned before the submission will be read and considered. The waiver should stipulate that the submission is made free of obligation on the firm's part, that the firm may be working on a similar project, and that the firm may use all or part of the idea and pay what it feels is reasonable. Exhibit 4.4 shows the terms and conditions used by one firm. You should develop your own, with the help of your legal department.
- Only after the signed waiver is received should the submission be reopened and considered.

Universities

University professors and researchers are a potential source of ideas. Scholars working in science, engineering, and medicine can offer a wealth of information on developments in their fields. They may lack an appreciation of the commercial potential of their work or the ability to commercialize it. To exploit this source, consider establishing contact with key researchers in your field at various campuses. For example, Emerson Electric kept a group of professors in engineering schools on retainer. Periodic meetings were held with this "braintrust" to ensure that Emerson people remained up to date with developments in the area.

Sometimes university researchers do develop products or processes with commercial applications. Innovation centers have been established on many campuses to help in the technology-transfer process. But their track record is not spectacular, largely because of a lack of the resources and skills needed to commercialize the product or process. An established firm can provide these resources. If you haven't done so, survey the universities in your area to locate the innovation centers. Get a list of available inventions and projects. Indicate that your firm is a potential partner in the commercialization of appropriate ventures.

Suggestion: A number of possible outside sources of new-product ideas have been identified. Not all will be applicable to your firm, but some will. With your idea person in place, the task becomes one of identifying the appropriate sources. Next, determine how best to gain access to the source — how to establish a "flow-line" between the source and your idea person. Start with the suggestions set out

Exhibit 4.4.   One Firm's Policy Statement on Idea Submissions

---

AMSCO POLICIES CONCERNING IDEAS SUBMITTED BY OTHER THAN AMSCO EMPLOYEES

The American Sterilizer Company receives many ideas and inventions submitted either without thought of remuneration, or without a statement of the terms on which the idea or invention is offered. We appreciate receiving ideas from other than AMSCO employees and assure such persons that their ideas will receive thorough and fair consideration.

Since we are continually carrying on a large research and development program, your submitted ideas may pertain to a subject already under consideration by us. Also, the ideas may have been previously known by us. For these and other reasons, all ideas disclosed to us must be with the understanding, and on condition, that we assume no obligations of any kind whatsoever relating to the submitted ideas.

To avoid unintentional misunderstanding, we are listing here the conditions under which we will consider suggested ideas. Suggestions as to the manner of submitting the ideas are also set forth hereinafter.

PATENTABLE IDEAS

Ideas submitted to AMSCO and believed to be patentable should be fully disclosed to enable AMSCO to determine its interest in acquiring patent rights therein. Such ideas may be submitted by one of the following three modes of procedure:

1. File an application for patent, wait until the patent is granted, and bring the patent to the attention of AMSCO.

2. File an application for patent and then submit a copy of the patent application to AMSCO for consideration. A patent attorney should be consulted before proceeding in this manner.

3. Make a written description and/or sketch of the invention, sign and date both the description and the sketch, and submit to AMSCO for consideration. A duplicate copy of each paper should be retained by the inventor. It is suggested that prior to submission, the idea be disclosed to a person capable of understanding the invention, and the signature of such person be affixed on the copies of description and sketch retained for record purposes. Rough sketches are acceptable and it is unnecessary that any particular phraseology be employed in the description. It is important, however, that the idea or invention be clearly described in such a manner that a person possessing normal skill in the art to which it relates can readily understand just what the inventor proposes to do and how he proposes to do it. It is helpful if the inventor points out just what he believes to be new, together with the advantages which he claims the invention has over known devices or processes. AMSCO wants every inventor to be fully *protected under the patent laws on ideas submitted to AMSCO*. If the procedure outlined herein is followed, there is no need for the parties to enter into a confidential relationship in negotiations pertaining to patentable inventions.

When considering inventions submitted, AMSCO assumes no obligation other than to state whether or not it is interested in acquiring rights in the invention. It is understood that an inventor, in submitting his idea to AMSCO, relies solely on such rights as he has or can obtain under the patent laws. If a submitted idea appears to be patentable and is of interest to AMSCO, negotiations for the sale or licensing of patent rights may be stated at the time of submission of the invention.

NON-PATENTABLE IDEAS

If a person has an idea relating to changes in the products, machines or manufacturing processes of AMSCO or the uses of AMSCO products, which may not be patentable because it involves merely the *skill of the artist*, as distinguished from *invention*, it should also be sent to the Chairman, Patent Review Committee. Since such ideas cannot be protected by patents, competitors are free to appropriate them. Consequently ideas of this type are normally of but nominal, if any value. AMSCO has a well-developed suggestion system throughout its plants and is constantly adopting improvements based on suggestions submitted by its engineers, workmen and other employees familiar with its work and products. If ideas of this type are accepted from other than AMSCO employees, AMSCO will decide what compensation, if any, will be paid for each disclosure, if used.

GENERAL CONDITIONS

1. AMSCO will not consider any idea or invention submitted on the basis that it will be treated confidential or secret. AMSCO will not become a party to a confidential disclosure or consider an idea or invention on the condition that if it is not found of interest, it shall be kept a secret. It may be necessary to refer such an idea or invention to several persons, both within and outside the AMSCO organization to determine the probable value thereof. While there is no intention of giving out data on submitted ideas, secrecy cannot be assured. Every reasonable effort will be exercised to avoid unnecessary disclosure of such ideas. Any waiver of the above-stated rule by AMSCO must be in writing and signed by an officer of the Company.

2. Occasionally, AMSCO has under consideration, or is developing, a process or product which anticipates, completely or in part, an idea submitted by an outside party. Since AMSCO must remain free to utilize the results of its own large investment in research and development, ideas relating to subjects either under consideration or previously explored must necessarily be rejected. Under these circumstances AMSCO must be placed under no obligation to disclose either the results of its research and development or its interest in or appraisal of the subject matter submitted. A statement that the company is not interested in a submitted idea may therefore not be construed as a presumption on the part of AMSCO that the person submitting the idea has a property right therein. With the exception of unpatentable ideas, negotiations are conducted on the basis that the patent rights, which have been or might thereafter be obtained in the name of persons not employed by AMSCO constitute the subject matter of the negotiations.

3. AMSCO will not consider an idea or invention on the condition that an agreement or contract be consummated prior to disclosure of the idea or invention.

SUBMISSION BY LETTER

Submission of ideas by letter is the most satisfactory mode of procedure. The party submitting the idea should in every instance keep complete copies of everything sent to AMSCO for consideration. If, after an idea has been submitted, a personal interview seems desirable, it may be arranged.

Letters submitting inventions or other ideas to AMSCO have received and will continue to receive, personal replies. A copy of this pamphlet is included with each such reply, and continuation of negotiations, upon receipt of this pamphlet, is on the basis of conditions herein stated.

Reprinted with permission from *Generating New Product Ideas*, report no. 546 (New York: The Conference Board, 1972).

above; once an attempt is made, you will probably discover other sources and more convenient or effective ways of reaching them. Then you will be on your way to having an effective external idea-generating system in place.

Getting Ideas from Inside the Company

How many people are employed in your company? Several dozen, a hundred, five thousand, more? It is not unrealistic to assume that each employee might have one new-product idea during his or her time at the company. Employees are an excellent potential sources of new-product ideas, but all too often they are not heard from. Why? Here are some frequently cited reasons:

- *Our employees aren't very creative.* This may be partly true, but what has your firm done lately to encourage or stimulate their creativity?
- *Our employees don't know what we're looking for, and submit nonsense ideas.* Again, only partly right. Maybe your employees should be told what the firm is looking for in the way of ideas. Not all ideas will be winners, of course. But if you discourage all ideas simply because many or even most are bad, you may be missing out on a handful of potentially great ideas.
- *Even if an employee did have a good idea, he or she wouldn't know what to do with it — where to send it for action.* This is a common problem. The appointment of an idea person, together with some in-house publicity, will overcome this objection.
- *People who do submit ideas get discouraged; they used to send in ideas, but never had a response.* The best way to kill internal idea-generation is to provide no feedback. The presence of an identifiable idea person will encourage employees to submit ideas; they will know that they can expect a response from that person.

In short, many of the barriers to internal idea-generation can be easily removed. Given the number of employees in your company and their potential for generating new-product ideas, isn't it worth a little time and effort to encourage them to submit their ideas?

In this part of the chapter we'll look at some specific steps that can be taken. We will begin with some fairly simple general approaches; later we will look at ways to stimulate ideas from two special groups — the sales force and the technical people.

Suggestion Schemes

Probably the least expensive way to solicit new-product ideas is to implement a suggestion scheme. Most firms have cost-savings suggestion schemes. Posters are mounted on bulletin boards throughout the office, technical, and manufacturing areas urging employees to submit their cost-saving ideas. Sometimes prizes or rewards are offered. But new-product suggestion schemes are rare. Management probably assumes that employees are able to generate cost-saving ideas, but are dullards when it comes to money-making new-product ideas. This is nonsense.

Try implementing a new-product suggestion scheme for your employees. There are several possible formats:

- an ongoing suggestion scheme in which employees fill out a suggestion form and drop it in a box or mail it to the idea person;
- a contest, perhaps an "Idea of the Month" contest, with prizes given every month or two for the best idea;
- a targeted effort, such as sending memos to employee groups specifically asking them for their new-product ideas.

The following tips will be useful in any suggestion scheme:[4]

- Publicize the scheme widely. Make sure that every employee is aware of the program. Use posters, the company magazine or newsletter, and meetings to generate enthusiasm at all levels. Some years ago, for example, Corning Glass Works held a new-product contest featuring "New Product Telegrams" mailed to each employee's home asking for new-product ideas. The campaign was bolstered by outdoor signs, posters, slogans, and bulletins in plants and offices. After five months, a total of 26,687 new-product ideas were received!
- Handle ideas promptly and provide feedback to the sender. You must be organized enough to conduct a quick yet proficient screening of the many ideas submitted. Your response must make it clear that the idea was carefully considered even though it was rejected. Send a letter or note (no form letters, please!) to the employee outlining the idea's strong and weak points and stating the reasons for rejection or acceptance. (In the next chapter we'll look more closely at idea screening, an essential second step in an idea-generating scheme.) One new-products manager in a 3M division actively solicits new-product ideas from employees. Every two weeks, his screening committee religiously sits down to review

recent submissions. A checklist of questions is used to rate ideas. The employee is guaranteed a response to his or her idea within three weeks of submission.

- Welcome all ideas, and don't expect them all to be winners. Never belittle any idea. Even though the idea may appear inane, the submitter should be thanked and encouraged. It may be necessary to sift through many impractical or unoriginal ideas to find one gem.

- Provide guidance and assistance. Try spelling out what makes an idea acceptable: the new-product mandate of the firm is a good place to begin, followed by the more important screening criteria. But don't be over-restrictive; you don't want the employees to prejudge their ideas. If possible, provide the name and phone number of a person with whom employees can discuss their ideas and obtain help in developing them for submission.

- Offer incentives. When ideas are good ones, employees should be rewarded. The rewards can be in the form of prizes — cash or gifts — for the best ideas of the month, for example. Some firms offer recognition as the reward. Tying financial rewards to profits from the product idea is impractical — the time lag is too long. The rewards should be immediate in order to stimulate more ideas. Of course, the rewards and the winners should be publicized, both for recognition of the individual and to create additional interest in the suggestion scheme. A formal presentation at an annual banquet adds a special touch.

- Manage the suggestion scheme. Management should set objectives; aim at generating a specified minimum number of ideas within a given period. The program should be an ongoing one, and should be given periodic publicity boosts to maintain employee interest. Finally, the efficacy of the program should be reviewed annually and necessary changes and improvements made promptly.

Creativity Methods

"Creativity methods" are techniques used with groups to enhance creativity and idea-generation. These methods include: brainstorming, synectics, reverse brainstorming, attribute listing, focus groups, and others. Each is claimed to yield improved idea-generation and each is a favorite of one consulting firm or another. Rather than provide a complete description of each method, let's take a close look at the most widely used method, brainstorming, and ways in which the technique can be used in your firm. Brainstorming is the easiest type of creativity session to run, and can be done in-house without using an outside consulting firm or a professional moderator.

The theory behind brainstorming is that, given the right environment, most people can be creative. Creativity is thought to be especially stimulated in a group environment, where one idea triggers another. The problem with most group meetings in a business setting is that the environment is totally wrong for any kind of creative thinking.

Consider for a moment the mental processes that occur in a typical business meeting. Two modes prevail — *ideation* and *evaluation*. A meeting is convened to discuss a problem; one person suggests a solution. That's the ideation mode at work: coming up with ideas, being creative. Immediately the rest of the group comments on the suggestion — that's the evaluation mode. The difficulty is that creativity suffers when one moves back and forth between the two modes. Usually, the evaluation mode dominates, at the expense of ideation.

For a brainstorming session to work effectively, the participants must remain in the ideation mode, even though the natural tendency is to jump immediately to the evaluation mode. What follows here is a blow-by-blow description of an actual brainstorming session at work. This example will show you how to set up and run an effective creative session. The example is taken from my own case files.

The Problem

A firm owned a relatively small, uneconomic nuclear-power plant that had been built in the 1960s as an experimental facility. The plant had been closed for some years, but was still costly to maintain. It was located in a remote area, had a large exclusion zone (no nearby population), had a ten-story concrete containment building, was on a river, and had a port facility and a rail siding. Management wanted proposals on what to do with the facility: "Surely there is some product or service that can be made using the facility, and at a profit!" exclaimed a frustrated company official. The project manager, who had been working on the problem for over a year, claimed to have investigated five ideas for the site, of which "only one and a half were at all decent."

A brainstorming session was proposed. Personnel from various groups within the firm — engineering, marketing, business development, and operations — were invited to the all-day session. (In some cases knowledgeable people from outside the company are invited to such sessions.) The fact that the group was multidisciplinary ensured that different points of view were expressed. The session was held away from the office to minimize distraction and interruptions.

Setting the Stage

The preliminaries involved introductions and small talk over coffee. Next, the meeting was convened. The project manager — the "client" in this case — explained the problem: the need to find a way to make profitable use of the disused plant. The moderator, in this case myself, explained what brainstorming was and what the rules were.

Each participant was handed a card and asked to write down his or her best idea for solving the problem. After a minute of this ideation, I collected the cards, picked one at random, and read the idea to the group. I then began to comment on the idea, with phrases such as "Yes, but..." and "Haven't we already looked at this one before?" and "The boss won't go for this one." The rest of the group jumped in and continued the attack on the idea. It was easy; we managers are very good at killing ideas. The critique was stopped after a minute or two. I then confessed that this was a demonstration of the two modes

Exhibit 4.5. Two Modes of Management Thinking

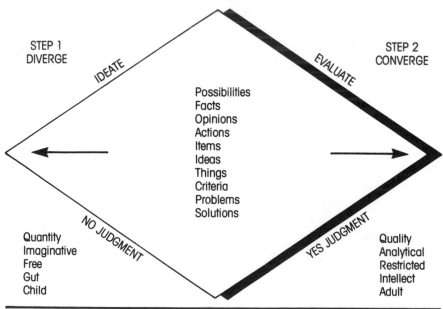

Reprinted with permission from Min Basadur, *Ideas, Creativity, and Innovation* (Cambridge, Mass.: Productivity Press, forthcoming).

of mental processes, ideation and evaluation, and that the best way to kill ideation is to shift to the more comfortable mode of evaluation.

The theory underlying brainstorming was then explained. In the ideation mode, we diverge: the focus is on creativity and on the quantity of ideas, and we behave like children. In the evaluation mode, we converge: the focus is on judging and on the quality of ideas, and we behave like adults. (Exhibit 4.5 graphically illustrates the two modes.) It is nearly impossible to operate in both modes at the same time: the "adult" evaluation mode will dominate. Any one of the "killer phrases" shown in Exhibit 4.6 will effectively put an end to a nascent idea. I showed the list to the participants in the session, saying, "Look, I've even made the job of killing ideas easy for you. You don't even have to say the words; just sing out a number, and ideation will grind to a halt."

The group was then told that for the morning session they were to operate strictly in the ideation mode. All participants were required to agree to the rules before the session began. Judgment was to be deferred. No judgmental response of any kind — words, facial expression, or body language — was permitted. Anyone expressing such a reaction would immediately be asked to leave.

The Session

I then explained that I would write ideas on the flip chart as the group called them out, and said "Let's go!" (A "scribe" was also present to record the ideas.) A moment of silence ensued, and then the

Exhibit 4.6. Killer Phrases

1. It's a good idea, but…	13. That's not our problem.
2. It's against company policy.	14. The boss won't go for it.
3. It's all right in theory.	15. The oldtimers won't use it.
4. Be practical.	16. It's too hard to administer.
5. It costs too much.	17. We've been doing it this way for a long time.
6. Don't start anything yet.	
7. It needs more study.	18. If it's such a good idea, why hasn't someone suggested it before?
8. It's not budgeted.	
9. It's not good enough.	19. You're ahead of the times.
10. It's not part of your job.	20. Let's discuss it.
11. Let's make a survey.	21. Let's form a committee.
12. Let's sit on it for a while.	22. We've never done it that way.
	23. Who else has tried it?

Reprinted with permission from Min Basadur, *Ideas, Creativity, and Innovation* (Cambridge, Mass.: Productivity Press, forthcoming).

ideas started to flow, slowly at first, and then with increasing momentum as participants began to relax in the supportive climate. Ideas were numbered, and as each flip-chart page was filled it was posted on the wall. After about an hour, the pace slowed down, but the session continued. Studies show that the quality of ideas does not diminish as a session proceeds; the last third of ideas generated is as good as the first or middle third even though the pace of ideation slows.[5]

The session was finally halted after three hours. The results were impressive: 92 ideas were generated, the majority of them original. (Compare this result with the project manager's five ideas in one year.) In subsequent sessions, the group shifted to the judgement mode, and the 92 ideas were evaluated, first on a set of "must criteria" (22 ideas survived this first cut), and then on a set of "should criteria," leaving six "best bets." Following a thorough analysis of the six best bets, the most promising solution to the nuclear-facility problem was found to be an original idea that had emerged rather late in the brainstorming session.

An Effective Tool

Properly implemented, brainstorming can be an extremely effective tool in generating new-product ideas. Remember the keys to an effective brainstorming session:

- Select a multidisciplinary group (10 to 15 people) from similar levels in the firm (avoid the senior executive-junior manager mix; the senior person tends to dominate, and the junior person tends to posture).
- Get away from the office. Set aside an adequate and uninterrupted period of time in a hotel suite or meeting room.
- Ensure that the theory and rules of brainstorming are explained and understood. Make use of the formats and exhibits contained in this chapter to clarify your explanations.
- The moderator should play a low-involvement, objective role, writing the ideas on a flip chart as they're called out and posting them so that they are easily visible. Appoint a "scribe" to record the ideas.

Suggestion: Try running a new-product brainstorming session at your firm. Invite a small multidisciplinary group. Explain the two modes of thinking, using the exhibits to make your point. Remember to act as a low-involvement moderator. You will be pleasantly surprised by the results of the session.

Special Groups in the Company

New-product contests, suggestion schemes, and creativity methods such as brainstorming can be opened to most or all employees. But two employee groups merit special attention: the sales force, because they are the front-line troops, the eyes and ears of the firm in the marketplace; and the technical groups, including R & D, engineering, and design, because they have access to potential new-product ideas from professional and technical sources. There are some specific steps you can take to stimulate idea-generation from both groups. For example, if you're reluctant to attempt a company-wide suggestion scheme or contest, try one targeted specifically at sales and technical employees. A smaller scheme is easy to manage and should yield a number of ideas.

Have your idea person attend meetings of sales and technical groups to explain the importance of new-product ideas, to define what is meant by a "good idea," and to describe what the idea-generator should do with his or her idea. When was the last time someone from your new-products group gave such a presentation at your national sales meeting? The mere presence of the idea person often will break down barriers to the flow of ideas. Have your people get to know some of the potential sources of ideas on a personal basis.

Salespeople spend much time on the road, and are often geographically remote from headquarters. Soliciting new-product ideas from them may be especially difficult. But of all people in the firm, they have the best opportunity to generate market-derived new-product ideas. True, your salespeople probably fill out call reports that indicate new-product opportunities. But chances are the call report sits in the sales manager's file cabinet, and reports on potential new-product opportunities never reach the new-products department.

Begin by designing a special report — one to be filled out only when a new-product opportunity is spotted. The opportunity may take the form of a customer complaint, problem, or request. This product-opportunity report should be sent not to the sales manager, but directly to the idea person (or whomever is charged with initiating new-product projects). A sample form is shown in Exhibit 4.7.

To complement this report, try setting up a new-products hot line for salespeople — a number they can call to submit a new-product idea. In addition to attending sales meetings, give the sales force other reminders. For example, a decal for their briefcases might read: "Got a new-product idea? Phone me, Jane Smith, on the New Products Hot Line at 555-0000." Every time the salesman opens or closes his briefcase he'll get the message. Finally, don't be afraid to provide appropriate rewards to salespeople for well-conceived new-product ideas.

Exhibit 4.7. New-Product Opportunity Form

SALES FORCE CALL FORM: NEW-PRODUCT OPPORTUNITY

Please complete this form each time you see an opportunity for a new product. Keep your eyes open for competitors' new products (report these!) and for problems with current products that could lead to a new product. Also, don't forget to note customer requests or ideas for a new product. Fill out this form and send it directly to _____

Describe the new product you propose: _____

_____

What's the reason for your idea? _____

_____

Describe the product's market (nature, size, competitors): _____

_____

Why will the product sell or succeed? _____

_____

Source of the idea (check as many as apply):

☐ A competitor's product. Specify competitor and give details below.

☐ A customer request. Specify customer and nature of request below.

☐ A customer complaint. Specify customer and nature of complaint below.

☐ Other. Specify and give details below.

Details: _____

_____

Your name: _____

Signature: _____

Office: _____ Phone: _____

Address: _____

Thanks for the idea. We'll report back to you as soon as we've had a chance to consider the idea.

Try running some creativity sessions with both sales and technical people. The best new-product ideas begin with a customer problem or need and involve a creative technical solution to the problem. Ideas that originate from the sales force alone tend to be market-derived; but the salesman, who simply doesn't know what is technically possible, can offer no imaginative solutions. Similarly, ideas originating from technical people — technology-push ideas — tend to be solutions looking for problems, a formula for failure. Why not try a combined group approach? At the next general sales meeting, ask to set up a brainstorming session or two with a select group made up of equal numbers of sales and technical people. The synergy between the two groups will likely yield more imaginative new-product ideas.

Management Must Lead

Creativity leading to idea-generation is a fragile thing. It can only flourish in a positive environment, and senior management is crucial to establishing that environment. Unfortunately, the beliefs, attitudes, and values of some senior managers are detrimental to a creative climate. For example, managers may fear new ideas, believe that good ideas come only "from the top," or be preoccupied with today's bottom line to the detriment of tomorrow's.[6] If top management is serious about implementing a successful and profitable new-product program, then it must lead the way. Senior managers should recognize that good new-product ideas represent the way out of "the product life-cycle trap."[7] Managers must come to realize that they are not the sole source of good new-product ideas, and they must learn to look to tomorrow and to worry less about the last quarter's results.

Suggestion: This chapter began with a challenge: to double, triple, or quadruple the number of new-product ideas that your firm will consider next year. The approach is straightforward. Identify the sources, establish an idea person, and set up lines of communication to facilitate the flow of ideas.

We have identified a number of potential outside sources of ideas and suggested ways to tap into them. Similarly, internal techniques — contests, suggestion schemes, and creativity methods — are effective in idea-generation inside the firm. Pick the sources and approaches that are most appropriate for your firm. Then design your own idea-generation system and list the ways in which it can be implemented. Exhibit 4.8 summarizes the actions you can take in developing your system.

If you carry out these steps, you are virtually certain of meeting the challenge of this chapter; next year at this time, you'll be looking at two, three, or four times the number of new-product ideas that you have now. Then you'll be faced with a different problem: how to reduce that number to the handful of best bets. This is the topic of chapter 5.

Exhibit 4.8. Twenty-five Steps Toward Generating New-Product Ideas

1. Establish a focal point in the company—an idea person.
2. Identify the possible sources of new-product ideas.
3. Use focus groups of customers or potential users to generate new-product ideas.
4. Set up a user panel that meets periodically to discuss problems or needs that might lead to new-product ideas.
5. Survey your customers.
6. Observe your customers as they use your product.
7. Install a customer hot line.
8. Maximize your sales and service staff access to and interaction with innovative users.
9. Hire sales and technical people who can recognize potential new products.
10. Promote your quest for new products to users by targeting likely innovators, defining the product desired, and providing a reward.
11. Routinely survey your competition.
12. Organize a trade-show visitation program.
13. Set up a clipping service for domestic and foreign trade publications.
14. Examine patent files and the *Official Gazette* regularly.
15. Use idea brokers and product-license brokers.
16. Attend product-licensing shows.
17. Visit your suppliers' labs and spend time with their technical people.
18. Set up a system to handle ideas submitted by private inventors in a legally sound fashion.
19. Visit key universities and researchers. Consider putting several key researchers on a retainer.
20. Set up a new-product idea suggestion scheme in your company.
21. Run a new-product idea contest complete with publicity and prizes for the best ideas.
22. Run several brainstorming sessions using in-house and outside people. Use the format described in this chapter.
23. Run a new-product contest targeted specifically at sales and technical people.
24. Establish lines of communication to the sales force. Use a new-product-opportunity call report, a telephone hot line, a reminder decal, and presentations at sales meetings.
25. Organize creativity sessions involving sales and technical people in the same session.

Picking the Right New-Product Project

Project Evaluation

Most new-product projects are losers. Either they fail commercially in the marketplace, or they are cancelled prior to product launch. Thus, picking the right new-product project for investment becomes a vitally important task. First, far more new-product ideas are conceived than there are resources in the company to commercialize them. Second, the great majority of the projects probably are not suitable for commercialization. While many commercial failures are the result of poor management and bad execution, many others are simply bad projects — they should have been killed at an early stage.

In an ideal new-product process, management would be able to identify the probable winners early in the game plan and allocate the firm's development resources to those projects. Failure rates would be kept low, misallocated resources kept to a minimum, and the return maximized. This chapter tackles the difficult question of how to select winning new-product projects. In it we focus on various approaches to project evaluation at the stages preceding product development.

Project evaluation is not as easy as it seems. Many firms have a mediocre track record in picking winners. There are several reasons for these difficulties. Project evaluation — making a GO/KILL decision on a project — ideally should reoccur at a number of points in the game plan. One difficulty is that the amount of information available increases as you move through the game plan. This suggests that the sophistication of successive evaluations, and even the methods of evaluation, should change from one stage to the next. For example, while financial analysis based on capital-budgeting techniques may be appropriate for project evaluation in the latter stages of the new-product process, such analysis is likely to do more harm than good at earlier stages.

Evaluation Techniques

A second difficulty in project selection is choosing an appropriate evaluation technique. Many techniques have been described in the literature. Several years ago, a researcher identified over 100 different techniques for screening and evaluating new-product projects.[1] Each was said by its proponents to be the method for choosing a successful project. A closer look at the various approaches reveals the wide differences in the amount and nature of information required for each particular approach.

Exhibit 5.1.   Allocating Resources Among Projects

Different projects, each with unique resource requirements; how do you allocate scarce resources?

A Multi-project Decision

A third difficulty with project evaluation is project isolation. Ideally, a GO/KILL decision on Project A should be made in the light of the costs, risks, and payoffs of Project B, Project C, and so on. In practice, this is often hard to do.

Imagine that you are a sailing instructor at a children's summer camp. Today is parents' visiting day. You've arranged a demonstration sailboat race. But unlike the normal boat race, the objective here is to have as many boats cross the finish line as possible (see Exhibit 5.1). You're in a canoe, supervising the junior sailors, and there's only so much help you can give. Just about every crew gets into trouble. The management problem you face is this: how do you allocate your limited resources among the various sailing crews? To add to the confusion, each boat is at a different stage of completion of the race, and each needs different amounts and types of help. Should you help Boat C, which is in the latter stages of the race but in some trouble? Or should you assist Boat D, which is starting out well, but may need help to keep on course? In the boat race, as in new-product management, a decision on one contender cannot be made without taking into consideration the needs of the others.

To the management scientist, the problem is one of constrained optimization under conditions of uncertainty: a multi-project, multi-stage decision model. But fancy optimization techniques usually don't work well in practice, and certainly not in new-product project selection. So we must resort to other, simpler techniques. Always recognize, however, that the true problem is one of resource allocation among a number of projects.

More Than Just a GO/KILL Decision

A further difficulty in project evaluation is the output. At minimum, project evaluation should yield a GO/KILL/HOLD decision. The GO decision means "Go to the next stage of the process, after which the project will be re-evaluated in the light of new and more concrete information." The KILL decision is self-evident: drop the project and cut your losses. The HOLD decision means "Put the project on the back burner; the evaluation is so far not positive enough to move to the next and more expensive stage, but is not negative enough to justify an outright KILL."

But an effective evaluation should do more than simply make the GO/KILL/HOLD decision. If the decision is GO, the evaluation should provide an answer to the question, "Go where?" That is, what steps should be undertaken next? What information must be obtained before the project is re-evaluated? In short, a good evaluation should identify possible "killer" variables and key areas of uncertainty and risk that must be investigated before the next major evaluation point in the project.

Summary

- Project evaluation must be undertaken at a number of points throughout the new-product project — not just at one or two points near the beginning or end of a project. The GO/KILL decision points are central to managing risk.
- There are many evaluation techniques, and the choice of the appropriate technique will depend on the type of project and its stage in the new-product process. The amount and quality of information available will limit the range of techniques that can be used effectively.
- The true problem is one of resource allocation — looking at one project in relation to the risks, costs, and payoffs of others at various stages of completion. This is not easy to do, and you may be forced to make simplifications; projects are often considered in relative isolation from one another. While it is necessary to evaluate each new-product idea on its own merits, other project opportunities and their status must always be borne in mind.
- An effective evaluation must do more than lead to the GO/KILL decision; it must also identify key areas of risk and uncertainty. The evaluation must answer the question, "What do we do next?" The evaluation influences the game plan for the next stage of the project.

Initial Screening: The First Critical Decision

The new-product process is characterized by a series of GO/KILL decision points designed to weed out poor projects and to focus resources on the best bets. At the birth of a new-product project, when it is little more than an idea, management must evaluate the available evidence to confirm the validity of the idea. More new-product projects are killed at the idea-screening stage than at any subsequent stage. It is at this stage that resources are first committed to a project: it is here that the project is born.

Screening Errors

A too-weak screening procedure fails to weed out obvious losers or misfits, and results in misallocation of scarce resources and the start of a creeping commitment to the wrong projects. Remember: the easiest point at which to kill a bad project is at the screening stage. After that point resources are committed, people become intensely involved, the project takes on a life of its own, and killing it becomes more and more difficult.

A too-rigid screening process results in many worthwhile projects being rejected, and is perhaps even more costly to the firm in terms of lost opportunities. At the idea stage, potential projects, particularly venturesome projects, are fragile things. An over-rigorous and conservative screening process that weeds out everything but the sure bets contributes to a "win the battle, lose the war" outcome for the new-product program.

The Requirements of a Good Screening Method

Many approaches to the idea-screening stage of the new-product process have been developed. When designing a screening process, bear in mind the following points.

- *The screening procedure is a tentative commitment in a sequential process.* The GO/KILL decision is only the first in a sequence of such decisions.[2] A GO decision is not irreversible, nor is it a decision to commit all the resources needed for the entire project. Rather, it can be viewed as a flickering green light for a project — as a decision not to reject. A commitment is made to spend a limited amount of time and money on the project — stage 2 — after which the project will undergo a more thorough evaluation with better information available. Using our earlier gambling analogy, the screening decision amounts to a decision to "buy a look" at the project.
- *The screening procedure must maintain a reasonable balance between errors of acceptance and errors of rejection.* The screening decision should not be over-conservative, accepting only the sure bets, nor should it lead to the dissipation of resources over a large number of unwarranted projects.
- *Screening is characterized by uncertainty of information and an absence of financial data.* The screening decision amounts to an investment decision that is made in the absence of reliable financial data.[3] The most accurate data in a project are not available

until the end of product development, just before commercialization.[4] Many of the activities of the new-product process, such as market studies, engineering studies and development work, test markets, and pilot production runs, are aimed at obtaining information. As the project advances, and as the fund of knowledge is augmented, project evaluations become more demanding. But at the initial screening stage, figures on projected sales, profit margins, and costs are little more than guesses (if they exist at all). This lack of financial data emphasizes the substantial difference between the methods needed for new-product screening and those employed for conventional commercial-investment decisions.[5]

- *Screening involves multiple objectives and therefore multiple evaluation criteria.* The criteria used in the screening decision should reflect the corporation's overall objectives, and in particular its goals for the new-product program. Obvious objectives are profitability and growth. But there could be others, including opening new opportunity windows, establishing acceptable risk levels, or complementing existing product lines. Not all of these objectives and selection criteria are quantifiable: for example, the desire to move into a specific new technology is not as easily quantified as a "20 per cent discounted cash flow (DCF) return" criterion. Nor are the criteria necessarily consistent with one another: there might arise a conflict between the "new technology" criterion (expensive, high-risk projects) and the "20 per cent return" objective. Finally, the evaluation criteria do not necessarily remain the same as the project progresses. As better data become available toward the commercialization phase, the decision criteria become better defined, and may even change over time.[6]

- *The screening method must be realistic and easy to use.* Screening must not entail so many simplifying assumptions that its output is no longer valid. Many operations-research screening tools fail on this point, largely because their simplifying assumptions render the method unrealistic. For example, resources are often assumed to be easily transferred between one project and the next, or between stages of the project, which is usually not true. The method must be easy to use in its data requirements, computational procedures, and interpretation of results.

No one screening method meets all the requirements outlined above. In the section that follows, we will review a variety of screening methods.

The four main approaches to initial screening include:

- benefit-measurement models;
- economic models;
- portfolio-selection models; and
- market-research models.[7]

Benefit-Measurement Models

Benefit-measurement models require a well-informed respondent or group to provide subjective information regarding characteristics of the project under consideration.[8] They typically avoid conventional economic data such as projected sales, profit margins, and costs, and rely on subjective assessments of fit with corporate objectives. Included in this category are checklists and scoring models. In the latter, ratings of project attributes are weighted and combined to yield a project score.

Benefit-measurement models recognize the lack of concrete financial data at the idea stage, and rely on subjective inputs only. But they treat the screening decision of a project in isolation, and do not take into consideration the impact of the project on overall resource allocation.

Economic Models

Economic models treat the idea-screening decision much like a conventional investment decision. Computation approaches, such as payback period, break-even analysis, return on investment (ROI), and DCF (capital budgeting) methods are used. To accommodate the uncertainty of data, probabilistic techniques, including Monte Carlo simulation and decision-tree analysis, are proposed.

At the idea-screening stage, however, economic approaches suffer because they require considerable financial data at a time when relatively little is known about the project. These models are usually considered more relevant for "known" projects (such as line extensions or product modifications — projects that are close to home, and for which relatively good financial data are available) or at later stages of the new-product project. Like benefit models, economic models consider each project in isolation, and do not deal with the overall resource-allocation problem.

Portfolio-Selection Models

Portfolio-selection models employ operations research constrained-optimization methods such as linear, integer, and dynamic programming. The objective is to develop a portfolio of new and existing projects to maximize an objective function (for example, expected profits) subject to resource-allocation constraints. These mathematical models require considerable data inputs, including financial data on all projects (both new and existing), timing information, resource availability by type, and probability of successful completion.

Market-Research Approaches

Market-research methods are usually used in connection with relatively simple consumer products, such as consumer goods. Market-research techniques assume that the sole criterion for moving ahead with a new-product project involves expected market acceptance and that technological and production issues are easily solvable. Given a market-based screening decision, it makes sense to use a variety of market-research techniques ranging from consumer panels and focus groups to perceptual and preference mapping. The danger, of course, is that this approach will produce a one-sided evaluation and that issues other than market acceptance of the product will be neglected.

Which Methods Are Used in Practice?

Which of the approaches to screening are popular in practice? Although conceptually appealing, portfolio approaches are used very infrequently. Studies done in North America and Europe show that management has a great aversion to these mathematical techniques, and for good reason.[9] The major obstacle is the amount of data required — information on the financial outlook, resource needs, timing, and probabilities of completion and success for all projects. Much of this information simply is not available. Portfolio approaches also provide an inadequate treatment of risk and uncertainty; they are unable to handle multiple and interrelated criteria; and they fail to recognize project interrelationships with respect to the payoffs of combined utilization of resources. Finally, managers perceive such techniques as being too difficult to understand and use.[10]

Problems with Economic Analysis

Economic models are often used as screening tools (see Exhibit 5.2).[11] They are familiar to managers, and they are accepted in other types of investment analysis in the firm. But they do have limited applications. Economic models require financial data as input. Someone must make estimates of expected sales in year 1, year 2, and so on; and estimates are required for selling prices, production costs, marketing expenses, and investment outlays. Often these variables are difficult to estimate, especially in the early stages of a project. And even when estimates are done, they tend to be inaccurate. More's study on firms' abilities to estimate expected new-product sales confirms that estimates were in error not by 10 per cent or 20 per cent, but by orders of magnitude.[12] Applying a financial screen to a high-risk or step-out project will tend to kill it.

The problem is not the project, which may well be a viable one; the problem is the application of the wrong evaluation technique. A

Exhibit 5.2.   Management Use of Screening Methods (104 firms)

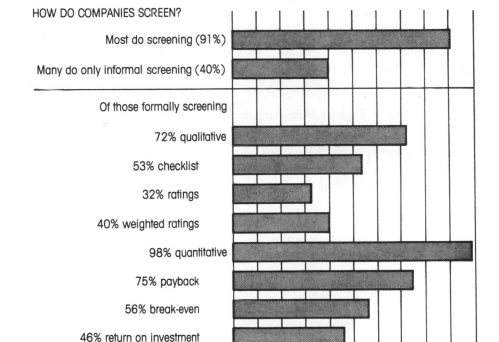

HOW DO COMPANIES SCREEN?

Most do screening (91%)

Many do only informal screening (40%)

Of those formally screening

72% qualitative

53% checklist

32% ratings

40% weighted ratings

98% quantitative

75% payback

56% break-even

46% return on investment

48% DCF (discounted cash flow)

Source: L.M. Katz, "New Product Screening Techniques," MBA dissertation, McGill University, 1974.

financial analysis that is done prematurely on a project will kill all but the sure bets, and will drive the firm into a conservative new-product program.

Remember that financial analysis involving economic models (payback, ROI, DCF, etc.) is a powerful and useful tool in project evaluation, provided it is used at the right time and for the appropriate project type. If used too soon, or used for the wrong projects, it can do much damage. Qualitative and nonfinancial considerations must also enter the decision to move ahead. Therefore, limit the use of financial evaluation to "known" projects — line extensions, product modifications, and the like — at the early stages. For more venturesome new products, avoid the use of financial techniques until a later stage in the game plan.

Problems with Market-Based Screening Methods

Market-research techniques also see some degree of use in screening new-product projects. The use of such techniques amounts to collapsing the first few stages of the new-product process — screening, preliminary assessment, and concept — into a single stage. For relatively simple new products, this may make sense. But often there are considerations other than market that must be evaluated prior to proceeding with a project. Before you move ahead with a market-research screen, the project should, at minimum, meet a set of "must" and "should" screening criteria.

The Benefits of Benefit Measurement

Benefit-measurement methods are generally recommended for new-product screening. Because only a tentative commitment is required at this early stage, and since available information tends to be limited, benefit methods are the most logical screening tool. Of all the benefit methods, the checklist and scoring model (based on a weighted checklist) appear most popular. The Conference Board reports that about 53 per cent of the firms studied use written guidelines or rules for project selection, usually in the form of checklists or scoring models.[13]

An investigation of 26 project-selection techniques, including scoring, financial, and portfolio methods, was undertaken by Souder.[14] The managers who took part in the study rated the portfolio models the highest in terms of realism, flexibility, and capability. (Note, however, that they are also the most difficult to use, and require data inputs beyond the capability of most managers.) Scoring models were rated the highest in terms of cost and ease of use. Souder concludes that scoring models are "highly suitable for preliminary screening decisions where only gross distinctions are required among projects."

Suggestion: Take a look at how new-product ideas are screened in your firm. Is screening recognized as a stage in the process at which a GO/KILL decision can be made? What method is used to make these early GO/KILL decisions? If you're not using any method at all, or if you're relying strictly on a financial analysis, chances are your new-product screening can be improved. Read on to see how you can use benefit-measurement methods to implement a screening method that works.

Benefit-Measurement Models

Benefit-measurement models are popular and are recommended by a number of experts. Let's take a closer look at various types of approaches useful in screening new-product projects. These approaches are all intended to integrate subjective inputs:

- comparative approaches;
- profile charts;
- benefit-contribution techniques (financial indices);
- simple checklists; and
- scoring models (weighted checklists).

Comparative Approaches

Comparative approaches include such methods as Q-sort, project ranking, paired comparisons, and successive comparisons. Each requires the respondents or evaluators to compare one proposal to another proposal or to some set of alternative proposals. The evaluator must specify which of the proposed new-product projects is preferred, and, in some methods, the strength of preferences. A set of project-benefit measurements is then computed by performing mathematical operations on the stated preferences.

The Q-sort method suggested by Souder is one of the simplest and most effective methods for rank-ordering a set of new-product proposals.[15] Q-sort combines the use of psychometric methods with controlled group interaction. Each member of the group is given a deck of cards, with each card bearing the name or description of one of the projects. He or she sorts and resorts the deck into five categories, from a "high" group to a "low" group, evaluating each project according to a pre-specified criterion. (The criterion could be, for example, expected profitability.) The anonymous evaluators' results are tallied on a chart and displayed to the entire group. The group is then given a period of time in which the results are debated informally. The procedure is repeated, again on an anonymous and individual basis, followed by another discussion period. By the third or fourth round, the group usually moves to consensus on the ranking of the projects on each criterion. The method is simple, easy to understand, and straightforward to implement; it provides for group discussion and debate, and moves the group toward agreement in a structured way.

Comparative methods such as Q-sort do have their limitations. Perhaps their weakest aspect is that evaluators must give an overall or global opinion on a project. Individual facets of each project — for example, size of market, fit with distribution channels, likelihood of technical success — are never directly compared and measured across projects. It is left to each evaluator to consider these individual elements consciously or unconsciously and to arrive somehow at a global assessment. This may be asking too much of some evaluators. Moreover, the group discussion may focus on a few facets of the project and overlook other key elements. A second problem is that no cut-off criterion is provided; projects are merely rank-ordered. It is conceivable that even those projects ranked highest will be mediocre choices in a field of poor ideas.

Profile Models

Profile models display a set of information-accumulation patterns over the entire new-product process.[16] The process is subdivided into a number of stages; at each stage the necessary information that must have been gathered by that point is specified — for example, information on market size, on distribution costs, or on development costs. For a given project, the evaluator simply indicates his or her confidence in the information about a particular factor available at this time. This profile of the project in question is then compared to the desired profile.

To develop these profile charts, a time scale is shown (see Exhibit 5.3), beginning with initial screening and ending with commercialization. Numbers from one to five indicate the level of confidence in information on each element as a function of the elapsed time of the project. That is, how good is the information on each element at this time? Exhibit 5.3 shows a project about one-third complete, with increasingly improving information on several key elements. Only a

Exhibit 5.3.   Profile Chart of a New-Product Project

ELEMENT	"GOODNESS OF INFORMATION" AT EACH EVALUATION			
Market need	-3	-1	+1	+3
Market size	+2	+2	+2	+2
Market trends	-5	-3	-3	-3
Competitive situation	-1	-1	+1	+1
Technical feasibility	+2	+2	+4	+3
Technical solution	-1	-1	+1	+3
Cost of development	-4	-2	+1	+4
Production costs	-2	-2	+2	+2
Production method	+4	+4	+5	+5
	Evaluation 1	Evaluation 2	Evaluation 3	Evaluation 4

handful of elements are shown; in practice there will be a longer list. This chart is then compared to a profile previously developed and agreed to by management. If information is substandard on any element, more and better data must be acquired before the project is allowed to proceed. Negative information, of course, may signal a KILL decision.

The profile-chart method is more a control tool than a GO/KILL decision tool. It is not focused strictly on the screening stage, but covers the entire new-product process from screening to launch. Its main use as a control device is its ability to prevent a project from advancing to the next stage when key information is missing. It also helps determine what action is needed to move the project toward successful completion.

One major weakness of the profile method is that the preparation of profiles, whether standard or for one project, is no easy task. First, the choice of the elements for consideration and the choice of what constitutes a desirable profile are arbitrary. Second, the evaluator must ask herself how she "feels" about the "goodness" of information for a given project. Finally, the method does not signal a GO or KILL decision, but merely GO or HOLD. As a screening device the profile chart has limited application, but as a control tool it has merit.

Benefit-Contribution Models

Benefit-contribution models require the evaluator to gauge the project's attractiveness in terms of its specific contribution to new-product or corporate objectives.[17] Since new-product objectives are usually financial, the benefit-contribution method typically amounts to measures of monetary return. Various economic indices — quick-and-dirty financial calculations — are usually used.

For the initial screening of projects, simple cost-and-benefit comparisons are employed. Such index methods require only straightforward financial data, and hence may be particularly suitable for screening certain types of projects. The attractiveness of a new-product proposal can be measured by using the following equation:

$$\text{attractiveness index} = \frac{\text{expected benefit}}{\text{cost}}$$

where:

$$\text{expected benefit} = \text{benefit} \times \text{probability}$$
$$\text{benefit} = \text{some simple measure of profits}$$
$$\text{cost} = \text{cost to execute the project}$$

For example, one major and highly successful industrial-products firm deliberately avoids the use of complex financial calculations in screening new-product proposals. Instead, it uses the following index:

$$\text{attractiveness} = \frac{\text{sales} \times \sqrt{\text{life}}}{\text{cost}}$$

where "sales" is the likely sales for a typical year once the product is on the market; "life" is the expected market life in years of the product (the square root of life is the firm's way of discounting the future); and "cost" is the cost of getting into the market (R & D, capital costs, launch costs, etc.).

Two arbitrary "hurdle points," X and Y, are used: if the index exceeds X, the project passes; if the index falls below Y, the project is killed; and if the index is between X and Y, more investigation is required. In conjunction with this financial index the firm also uses a scoring-model approach to assess the qualitative merits of the project.

A financial-index method is a gross simplification of a rigorous financial analysis. But at the screening stage, only gross distinctions between good projects and sure losers are required. The method has the advantage of not requiring detailed financial data, and thus it suits the screening stage.

Checklist Methods

Perhaps the simplest approach to screening new-product ideas is the checklist method. This approach can be likened to the question-naires that follow magazine articles, for example, on "A Happy Marriage." At the end of the article is a list of 20 questions. If you answer more than 10 "no's" out of 20, it's time to see your divorce lawyer. These types of diagnostic questionnaires are found in many fields, from medicine to personal planning. They are reasonably accurate in terms of diagnosing some ailment or problem.

How are checklists developed? A group of experts constructs a list of questions which they believe are useful discriminators in predict-ing or diagnosing a situation. A cut-off score is established — how many "yes" or "no" answers it takes to indicate the existence of a problem.

Checklists work well in new-product idea screening. "Must" ques-tions often appear at the top of the checklist — questions that must yield a "yes" answer for the idea to pass. If the "must" criteria are met, then we move to a set of "should" questions that measure whether or not the project possesses some desirable features. While not every "should" question requires a "yes" answer, a good project will have a certain minimum number of "yes's." The "must" criteria typically relate to strategic and resource questions: Does the project fit the new-product mandate of the company? Does the company have the necessary resources to undertake the project? The "should" criteria follow, describing the relative attractiveness of the project: Is the market large? Is it growing? Can the sales force easily handle the product?

In using the checklist, a project is presented to a group of evalu-ators — people in the company who possess knowledge about certain facets of the project, but who are not necessarily intimately involved in it. A multidisciplinary group is often the most successful in the evaluation task. Following the project briefing, the evaluators answer a set of questions rating the project on each element ("yes/no" or "favorable/unfavorable"), typically using a questionnaire. The answers are tallied and a profile and "score" for the project determined. A suitable pattern of responses (for example, a pre-specified number of "yes" replies) signals a GO decision.

Checklist methods offer an attractive approach to screening. Implementation is straightforward; a number of criteria are considered; the list ensures that vital considerations are not overlooked; and all projects are subjected to the same criteria. There are some problems with checklists, however. The choice of items for the list is arbitrary — they represent the compiler's best guesses about what factors are important to consider in evaluating a project. Some elements are likely to be more important than others, and the checklist method does not provide for weighting. The issue of what constitutes "an acceptable pattern of responses" remains a difficult one. Finally, the inputs or answers to the questions by evaluators are highly subjective.[18]

Scoring Models

An improved version of the checklist is the scoring model. In a scoring model, elements or aspects of a project are rated on, for example, a scale of zero to ten that reflects shades or degrees of "goodness" rather than on a "yes/no" response. Second, the questions themselves are weighted — a recognition that some questions or elements are more important than others.

The scoring method consists of two parts. Projects are first subjected to a set of "must" criteria in checklist form. These are inexpensive and easy questions to answer, and weed out obviously unsuitable projects. The remaining projects are then subjected to scoring. As with the checklist approach, a group of evaluators is chosen. The evaluators are briefed on the project; then, independently of one another, they rate each of a number of elements of the project. The scores are tallied, and a mean score for each element is computed by taking the arithmetical average across all evaluators for each question. The score (or mean score) for each question is multiplied by the weighting factor for that question, and summed across questions to yield a project score.

Let's look at a simplified version of a scoring model. Assume that three "should" criteria have been identified (in practice, 20 to 50 criteria are used). The three criteria are:

- how closely the needs of the project fit the resources of the firm — the level of synergy;
- how attractive the market is: size, growth, lack of competition, etc.;
- the product's differential advantage — is it superior to competitive products on the market?

Assume that management believes that the last item, product advantage, is the most important and is weighted at 10. Market attractiveness is next, at 7. Synergy is less important, and is given a 3. After the model is developed, four potential projects are reviewed. Two are rejected on the "must criteria" during a group discussion. The remaining two projects are then scored on the three "should" criteria.

Project A is a "close to home" product, and fits the company well: it can be made in your plant, sold by your sales force, and developed by your firm's product-development group. Score it 10 out of 10 on the synergy criterion. But its market is only so-so: score it 6 on market attractiveness. The product itself is fairly similar to competitive products, and rates only a 2 out of 10 on product advantage. Project B, in contrast, has a significant product advantage, and the raters give it a 10 out of 10 on that element. The market is slightly more attractive than A, and rates a 7. But the fit with the company scores only 3 out of 10.

Exhibit 5.4 shows how these two projects look using a scoring model. The first column shows the importance weights, which are decided as the model is developed, and remain fixed from project to project. Note that the sum of these "importances" is 20, so that a project could obtain a maximum score of 200 if it scored 10 out of 10

Exhibit 5.4.   Sample Results of Scoring Model Evaluations

Element	Importance weighting	Scores		Scores x importance	
		Project A	Project B	Project A	Project B
Synergy or fit	3	10	3	30	9
Market attractiveness	7	6	7	42	49
Product advantage	10	2	10	20	100
Totals	20			92	158

on all three questions. The next two columns show the scores for the two projects. The last two columns show the results of multiplying these scores by the weights.

What have we learned from this exercise? First, project B looks better than project A. Project B's total score is 158 versus only 92 for A. If you were considering several dozen projects, you would be able to rank-order them from best to worst. Second, project A's score is 92 out of a maximum of 200, or a score of 46 per cent. Project B scored 158 out of 200, or 79 per cent. Many firms use a score of 50 or 60 per cent as a cut-off, according to which project A would fail and project B would pass. In effect, the scoring model has enabled GO/KILL decisions to be made for these projects.

Third, the scoring model provides insights into the projects. Why did A do so badly? Can anything be done about it? Why did B do so well? Are you sure that its positive features were assessed accurately? Exhibit 5.4 shows that the factor that "killed" project A was product advantage — it scored only 2 out of 10. Perhaps you can improve the product advantage by building some new features or benefits into the product. If you can't, the project will remain dead.

Now consider project B. Its main driving factor was product advantage: of its 158-point score, 100 points came from that one element. Perhaps the next step in project B will be to verify the product's potential strength in the marketplace through an end-user market study of relative advantage.

Of the early screening methods, scoring models appear to fare the best in practice. They reduce the complex problem of making a GO/KILL decision on a project to a manageable number of specific questions. As with checklists, each project is subjected to assessment on a complete set of criteria, ensuring that critical issues aren't overlooked, as so often happens in unstructured discussion meetings. The method forces managers to consider the project in greater depth, and provides a forum for discussion. Unlike the checklist approach, scoring models recognize that some questions are more important than others, and individual ratings can be combined into a single project score.

In spite of their popularity, there are problems with scoring models.[19] They rely on the subjective opinion of managers, and the input data may or may not be completely reliable. But at the screening stage, often the only "data" available are management opinions. Most firms use multiple evaluators, together with confidence scores, to heighten the reliability of inputs. The premise is that the average decision-maker is optimal. The dilemma is that the likelihood of any one evaluator being "average" is about zero — hence the desire for multiple inputs.

Another criticism of scoring models concerns the arbitrariness of questions and the weights assigned to those questions. The question of the cut-off point — why not use a cut-off score of 50 instead of 60? — also is a contentious issue, as is the real meaning of the "project score." These criticisms, as we shall see later, have been largely overcome in recent years.

The Payoffs

Despite its shortcomings, a scoring model, in conjunction with a checklist of "must criteria," is the most effective way to screen new-product projects. The method helps to render a highly judgmental decision somewhat less subjective; it systematizes the review of projects; it focuses attention on the most relevant issues; it forces management to state goals and objectives clearly; it is easy to understand and use; and it is generally applicable across a broad range of situations and project types.[20]

A Viable Approach to Screening

We've had a brief look at various approaches to screening new products. Now your task is to develop an appropriate screening method for your firm. Remember that screening is a culling process. The objective is to subject projects initially to simple, easy-to-ask questions; in this way you pare down the list of projects so that only a few projects will be subjected to a more rigorous screening. My own experience and the findings of experts confirm that a combination of various screening methods is best.

The First Pass

The "first pass" or cut involves the project's meeting the "must criteria." This elimination is best handled by a checklist. Remember, these are the criteria that must be met by every project. The answers to the questions must be an unequivocal "yes." If the answer is "no," the project will be killed. The "must" questions are usually few and simple to answer; they serve the purpose of rejecting the obvious misfits. A sample set of "must criteria" is shown in Exhibit 5.5. These types of questions relate to three basic needs:

- Mandate fit: is the project within the-new product mandate of the firm? (This question, of course, presupposes that you've defined what the mandate of the firm is!)
- Feasibility: is the project feasible with available or obtainable resources of the firm? If it's not feasible — because it's too big a project for you, outside your technical capabilities, or simply too farfetched an idea — kill it here and now.
- Other requirements: these are firm-specific, and may include, for example, maximum or minimum market and/or project size; the existence of a product champion; or fit with a top-priority area for the firm.

Exhibit 5.5. A typical list of must criteria

	YES	NO
MANDATE FIT		
1. Does the proposed new product fit within a product area or category that is within the new-product mandate of our company?	___	___
2. Is the new product's market within the market boundaries defined by our firm for new products?	___	___
3. Is the technology required for the development of the product within the technology areas defined by our firm for new products?	___	___
4. Is the nature of the manufacturing process within the boundaries defined by our firm for new products?	___	___
FEASIBILITY		
1. Is it technologically feasible to design and develop the product?	___	___
2. Does our firm have the necessary resources to design and develop the product (if not, can these be readily acquired)?	___	___
3. Is it technologically feasible to manufacture the product?	___	___
4. Do we have the necessary resources to manufacture the product (if not, can these be readily acquired)?	___	___
5. Is there a market for the product and can it be reached?	___	___
6. Do we have the necessary sales, distribution, and service resources to market and service the product (if not, can these be readily acquired)?	___	___
OTHER "MUST" REQUIREMENTS		
1. Is the product's potential market at least $X million?	___	___
2. Will the expected sales of the product be at least $Y million (or Z million units)?	___	___
3. Is there an identified product champion — a person who has volunteered to push and support this project?	___	___

The Second Pass

The "second pass" or cut focuses on project attractiveness. Given that the project fits within the mandate of the firm, and given that it is feasible to design and manufacture, is it an attractive one to do? Unlike the "must criteria," attractiveness is a matter of degree and is also more difficult to assess. This is an area in which a scoring model is most useful. The scoring model is used to assess a project against ideal or desirable characteristics — the "should criteria." These "should criteria" typically fall into one of four categories:

- Product advantage: for example, the product should
    - offer unique benefits to end-users;
    - incorporate unique features and attributes not found in competitive products;
    - be lower-priced than competitive products;
    - be of higher quality than competitive products.

- Market attractiveness: for example, the market should
    - be a large one;
    - be a fast-growing one;
    - have no dominant competitors;
    - have no or little price competition;
    - have significant long-term potential.

- Synergy or fit with the firm: for example, the product should
    - be able to be sold by our existing sales force;
    - be able to be made in our plant;
    - be able to be developed by our engineering or R & D team in-house;
    - be sold to our present customers.

- Company-specific items: for example, the product should
    - help to even out production schedules;
    - open a window on new market opportunities;
    - make use of an underutilized technical skill or group.

These are only sample criteria. Later you will learn how to develop a list of criteria and a scoring model for your firm. Appendix A shows a commercially available scoring model, NewProd, already in use in a number of firms.[21]

The Third Pass

The "third pass" also focuses on relative attractiveness, but with a stronger financial orientation. If reliable data on expected sales, costs, prices, etc. are available, then a financial analysis is called for. Only those projects that are similar to a firm's existing products are likely to be blessed with good financial data, however. Examples of these might include minor modifications, extensions, and updates. But most new-product projects suffer from a lack of reliable financial data at the idea-screening stage, and any financial analysis done at that stage should be a relatively simple one. Remember the caution against a premature application of a financial analysis; it is certain to kill all but the sure bets.

If some financial data are available at this stage, a financial index, such as that shown in appendix B, or a simple payback period calculation is recommended.

Developing the "Must" and "Should" Criteria

How does one go about developing a checklist and a scoring model to handle the first two passes in screening? These are two approaches, and both have their merits.

Management Consensus

One approach to developing a scoring model is the management-consensus method.[22] This method can best be illustrated by a hypothetical example. An all-day session is convened for the purpose of developing a screening tool. Present are managers who are involved in screening and who have had past new-product experience. The session begins with an open discussion; the facilitator asks that all the "must criteria" be identified. A free-wheeling debate ensues, and a list of criteria is developed and agreed upon. These "must criteria" usually make up a fairly short list (refer to Exhibit 5.5).

Next, the discussion moves to the "should criteria." The facilitator asks participants to identify those characteristics of an ideal project that are known at the outset. (For example, a high return on investment is certainly a desirable feature of a project, but the ROI is not likely to be known with any reliability at the early stages.) Again, a free-wheeling discussion ensues, and the group has little difficulty in producing a list of 50 characteristics.

After those items are rewritten and clarified, the group is asked to consider the importance of each characteristic. Each manager independently assigns an importance score from zero to 10 to each item. The scores are then collected, tallied, and displayed anonymously on an overhead projector. Discussion takes place on each question, focusing on those where strong differences exist. The facilitator then instructs participants to go through the importance-scoring exercise again. The process is repeated, and by the third round the group has reached consensus.

At this point, the management group has identified the "must criteria," developed a comprehensive list of "should criteria," and agreed on importance weights for the "should criteria." The scoring model has been developed and agreed to by the very managers who must use it, thus ensuring a high degree of commitment. An appropriate cut-off criterion can be established and the model validated by running past company projects, both failures and successes, through the model.

Review of Past Projects

The second approach to developing a scoring model is more analytical, more expensive, and more powerful. It requires a fairly large sample of past projects for its raw material, and its use is thus restricted to larger firms.

As in our previous example, managers meet to identify a list of "must criteria" and a set of possible "should criteria." But management opinion ends there. Next, a sample of past commercial successes and failures is identified. People who are familiar with the projects are then asked to rate them on the "should criteria" — to conduct an after-the-fact analysis or post mortem. Up to 10 evaluators per project are used to obtain these ratings in firms where this method is used. The relative commercial success or failure of each project is also rated — the degree to which the product exceeded or fell short of the minimal acceptable profitability level for its type of investment. With a large sample of projects, and using appropriate statistical methods (such as factor analysis and multiple-regression analysis), a success equation can be generated which relates the degree of success to the various "should criteria." The coefficients of this equation become the weights applied to each criterion.

In short, this analytical approach simply relies on past experiences to derive a predictive equation or model that can be used to rate future projects. This method was used to develop the NewProd model reproduced in appendix A.

Using a Screening System

A number of firms use new-product screening systems such as the ones described above to sharpen their new-product decisions early in the process. In use, the screening procedure is relatively straightforward, and can be summarized in a few simple steps.

First, formal project proposals or ideas are submitted in writing. As much information as possible should be given in the submission, including a description of the product, its use, its intended market, its advantages and features, and how the company might handle the development, production, and marketing of the product.

First Pass: The "Must" Criteria

Typically, this bank or list of ideas is first evaluated by a screening committee which subjects the list of ideas to the "must criteria." These "must" questions, in checklist form, are discussed for each proposal, and "yes/no" answers given. Most proposals will likely fail at this first cut, thus easing the workload on the decision-making group.

Second Pass: The "Should" Criteria

If the project passes, it is then rated on a scoring model. Usually six to 10 evalutors are chosen. Some may be involved in the project; others are chosen because they are well informed on the topic. A multidisciplinary team whose members come from R & D, marketing, production, product management engineering, sales, and finance seems to work best.

A Briefing Session

A briefing session is held, and the proposed project is presented to evaluators. Each evaluator then retires and rates the project using the scoring-model questionnaire. In the typical system, items or questions are presented in statement form, and the evaluator indicates whether the statement describes the project or not (see appendix A for examples). In some systems, the evaluator also indicates his or her confidence in each of the answers. If a financial index is employed, it is built into the screening system here: evaluators are asked for their estimates of sales and costs as an adjunct to the scoring model questionnaire.

The questionnaires are then collected and the data analyzed. (This data processing can best be handled by a relatively simple computer program; some commercially available models feature extensive software packages.)

Debriefing

Next, a debriefing session is held. This session is probably the most valuable part of the screening process. The team of evaluators meets to review one another's input. Any areas of disagreement usually are readily identified. For example, in one evaluation of an industrial chemical, two of the evaluators assumed the existence of a positioning strategy and a niche target market that avoided a head-on confrontation with major competitors; the other three evaluators rated the project assuming a nose-to-nose positioning strategy with competitors. The input ratings and prognosis for the project of these two groups were quite different. The reasons for these differences became quickly apparent, however, and a discussion on positioning strategy ensued. The product was eventually developed and launched, targeted at the niche segment, and, as predicted, became a success.

Pros and Cons

The project's score and its strong and weak points are discussed in the debriefing. Finally, a GO/KILL decision is reached. If the decision is GO, a course of action is mapped out. The debriefing session is particularly useful in this respect; by highlighting areas of disagreement and uncertainty and "killer variables," the action requirements become obvious. Some typical questions are addressed: Why was the project rated so positively (or so negatively)? What are its good points? Are we sure about them? What are its bad points — the "killer variables"? Can we do anything to improve those negative factors?

A Course of Action

This type of impact analysis often results in the correction of a highly negative feature of a proposed project or the confirmation of a positive feature. For example, in the evaluation of a new building material, the product fared relatively well on most points — a good fit with the company, a solid and growing market, and relatively weak competition. The product itself was rated negatively, however. It didn't offer the customers any more benefits than the product they were already using. This one factor proved to be particularly damaging to the project's overall assessment. Discussions ensued, and the R & D manager confessed that, in proposing the project, his goal was simply to develop a product equal to that of the competition. The outcome of the debriefing meeting was a recognition that product superiority was essential to the product's success. Two tasks were immediately assigned: an end-user interview to identify weaknesses in competitive products as perceived by potential customers; and in-house creativity sessions (brainstorming) that focused on ways in which the proposed new product could be significantly improved so as to outshine its competition. The output of a well-conceived screening system is not only a GO/KILL decision, but an indication of what needs to be done next when the GO signal is given.

Tracking the Project

This chapter has described a proved and effective method for product selection involving three culling steps: "must criteria," "should criteria," and a quick financial evaluation using, for example, a financial index.

Although this project-selection method has been positioned at the initial screening stage, it is clear from our game-plan model that many more GO/KILL decisions must be made as the project progresses from the idea stage through to launch. A number of those decisions must be made prior to the product-development stage. What evaluation techniques should be used at these subsequent stages? Until you have sufficient reliable financial information to permit a detailed financial analysis, the recommendation is that you continue to "track" or evaluate the project using the scoring model outlined above in combination with a relatively simple financial calculation. As more and better information becomes available, successive screens using a combination approach will become more and more valid. Of course, each successive evaluation can be compared to the one that preceded it: is the project looking better or worse than the last time we evaluated it?

Project Evaluation: A Final Thought

It is impossible to remove all the guesswork and risk from new products. We must always deal with the future, and hence with uncertainty. Sharper project evaluation is possible, however, and new-product screening has been identified as one area in which some firms have been doing a much better job in recent years.

Suggestion: Take a hard look at how new-product projects are screened at the early stages in your firm. Use some specific case histories as illustrations. If your firm is typical, chances are that this process can be made more effective. If you identify the screening stage as less than excellent, why not move toward the development of a systematic screening method, following the steps outlined here? Get a management group together. Develop a list of "must criteria." Move on to the "should criteria," using either a management-consensus approach or a historical review to assign weights and define the cut-off criterion. Then implement a procedure for getting screening done for each project: set up a screening committee to handle the "first pass" and evaluation teams, complete with briefing sessions, questionnaires, and debriefing sessions, to gauge the "should criteria" or project attractiveness.

Those firms that have made the effort — and there are a number that have carefully designed and implemented new-product screening methods using the approaches outlined above — are now reaping the benefits. They're seeing better project-selection decisions and a much clearer picture of what must be done to turn the project into a winner.

Defining the Product Concept and Specifying the Product

It's What's Up Front That Counts

If you've followed the suggestions given in chapter 4, you've generated a number of new-product ideas — more than you might have imagined possible. And you've pared these many ideas down to a subset of "best bets" using the screening approaches outlined in chapter 5. Now it's time to start spending time, money, and effort on those best bets.

But before you turn your scientists, engineers, and designers loose on the project, there are several up-front or pre-development activities that are critical to success. We know these "homework" activities are critical: studies of better performers in the United States and Japan (the Japanese spend, on average, 25 per cent of their total new-product budgets before the product-development stage) and empirical observation of what goes wrong in many new-product projects all point to the need for more emphasis on the planning function.

Let us look at the game plan again, focusing this time on the pre-development stages (see Exhibit 6.1). Looking ahead, the product-development stage is typically the big-money stage; this is where approximately 45 per cent of your project spending will take place. By the time you're into product development, you should be fairly strongly committed to the project and to the direction it will take. By this stage the die has been cast!

The Key Pre-Development Decisions

Before moving into the product-development stage, a number of key decisions must be made. One obvious decision is the commitment to the project — you must be sufficiently convinced of its chances for success to invest in its development. Remember that by the time

Exhibit 6.1.   The Up-Front Stages in the Game Plan

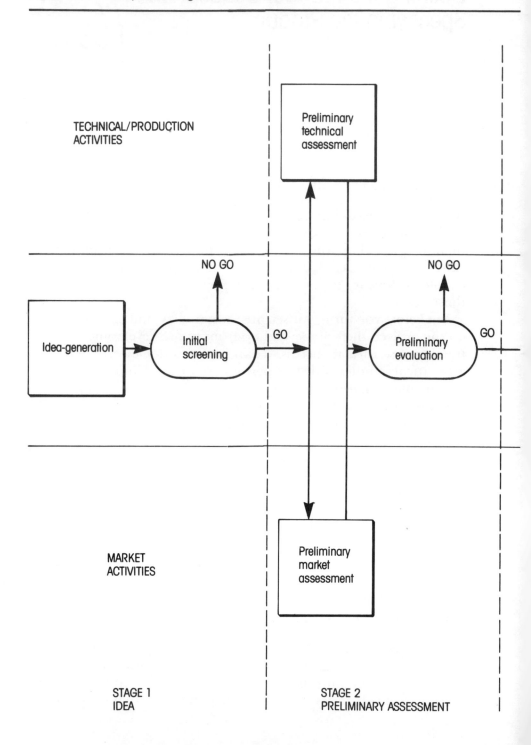

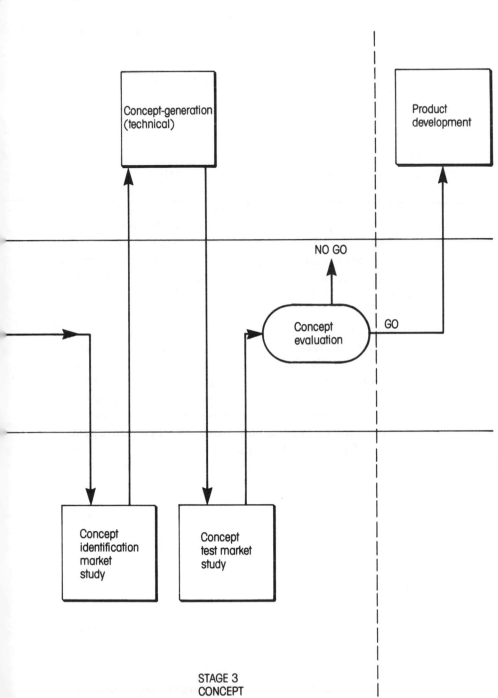

STAGE 3
CONCEPT

Adapted with permission from R.G. Cooper, "A Process Model for Industrial New Product
Development," *IEEE Trans. Engineering Management* EM30 (Feb. 1983): 2–11. Copyright ©1984 IEEE.

product development is underway, it is difficult and costly to kill the project; the decision to move into development is a critical one. Another decision (or set of decisions) focuses on the direction of the project. A common complaint by R & D groups charged with carrying out this product development phase is that there is "no target" or a "moving target."

In a no-target situation, the R & D group is not given a clear definition of what the product is, who it is aimed at, and what the customer's needs or requirements are. No matter how talented the scientist or designer is, he or she will have little success in developing a product in a virtual information vacuum.

The "moving target" situation is even worse. In this situation, the product developer thinks that he or she has a target — that the marketing or business development departments have defined the customer and product requirements. Their definitions were poorly conceived, however. The unsuspecting scientist or designer pushes ahead and develops the product as requested. He or she later discovers that the product requirements (and sometimes even the customer) have changed, simply because a sloppy job of project definition was done in the first place.

Many companies are guilty of this "moving target" problem. Recently I came across a classic example in a major U.S. cigarette company. The corporate marketing department had identified a target niche: the upwardly mobile young, black, male smoker. A menthol or flavored product was envisaged. The product-development department responded to the request. After conducting consumer research on the target segment, they developed the appropriate product. Roughly a year later the product was ready. In the meantime, the corporate marketing department had changed its mind. The product was repositioned and launched as a menthol cigarette aimed at white, middle-class, middle-aged women. The product, of course, was wrong for that market. Not surprisingly, it failed — another victim of a moving target.

Defining the Project

Before moving into full-scale product development, you must make sure that the project is worth proceeding with; equally important, you must define the target. But what is meant by "the target"? Answering the following questions will help you to clarify the target.

- Who is the customer? Describe the intended target market in detail.

- How will the product be positioned? Be clear about how the product will be perceived and differentiated from competitive products in the market.
- What benefits will the product deliver to the customer? These must be spelled out in unambiguous language.
- What are the product requirements? What does the customer need, want, or prefer in terms of product features, attributes, and specifications?

These four elements are the leading edge of the marketing strategy — the topic of chapter 8. But they're also critical to the product-development stage, and must be decided at this point in the game plan.

The Protocol: Getting Agreement

The four questions set out above must be answered, and the answers must be agreed upon by all the people and departments involved in the project. This agreement Crawford calls the "protocol."[1] In a new products context, Crawford views the protocol as a written statement, defining the product and its benefits, which is agreed upon prior to the commencement of product development. The "signatories" to the protocol are the departments of marketing, R & D, engineering, manufacturing, and any other involved parties.

Do not confuse the terms "protocol" and "prototype." A prototype, a familiar term to managers, is defined as "something that serves to illustrate the typical qualities of an item; a model." It is an early or first attempt at creating the physical product, and it is a *result of* the product-development stage. In contrast, the protocol *defines* the product-development stage and precedes that stage.

Defining a Winner

The final component of the protocol or project-definition step is a winning definition of the project. Merely going through the motions of defining the target market, the positioning strategy, the benefits of the product, and the product requirements isn't enough. Even gaining all-party agreement and a protocol document "signed" by all concerned won't suffice to create a winner. It is possible to execute these steps diligently, only to arrive at a broad and superficial definition of the market, an unimaginative positioning strategy, a ho-hum listing of product benefits, and a set of product requirements that is rather ordinary. Something is clearly missing. The key is to come up with a winning definition of the product and the market — a definition of a product that is unique and superior in the eyes of the customer.

The up-front steps that precede the product-development stage are crucial to creating this winning definition. During these first few stages of the process — stages 2 and 3 in our game plan — safeguards can be built in to ensure that a winning strategy or protocol is the result. These safeguards, which will enable a fuzzy new-product idea to be shaped, molded, and modified to create a winner, are the topic of this chapter. These up-front activities are vital steps; they can make the difference between success or failure; and they define the project's strategy or target. Sadly, in too many firms they are the weakest elements of the game plan.

Suggestion: How well defined are your new-product projects before they move into the product-development stage? Do you suffer from the "no target" or "moving target" maladies? If so, consider building in a protocol step just before serious development work begins. Even if the product is defined long before development begins, is the definition a winning definition? What steps have been taken to ensure a product that delivers superior benefits? Read on to see what actions can be built into your game plan to move toward a winning product definition.

Preliminary Assessment

The preliminary assessment is the first of the up-front stages. It really consists of three separate assessments:

- a preliminary market assessment;
- a preliminary technical assessment; and
- a preliminary project evaluation.

The Preliminary Market Assessment

The purpose of a preliminary market assessment is to qualify the project in a relatively inexpensive way. The key objectives of such a study are, typically,

- to obtain a rough estimate of market size and potential;
- to gain an indication of whether the product, as envisaged, has any hope of selling, and how well it might sell;

- to obtain insight into possible target markets, product benefits, product requirements, and pricing, distribution, promotion, and sales strategies.

Remember that there is a 50-50 chance that any project will be killed at the end of the preliminary assessment stage. The GO decision at the initial screening was only a flickering green light, not a strong commitment. The odds preclude forging ahead with an expensive "scientific" market-research study at this point. An informal investigation aimed at gaining insights and estimates rather than definitive answers is more appropriate. The objective is to gain enough information so that you can make a better decision about pushing the project ahead to the next and more expensive stage.

Some firms place a tight financial and time limitation on the entire preliminary-assessment stage. Approval may be given, for example, to spend no more than $5,000 and 10 man-days on researching the market and the product; following that period the project will be re-reviewed. What can be done in such a short time and with so little money? Depending on the project, a surprising amount! Here are some illustrations.

Secondary Data

There are very few markets or products about which nothing has been written. Several well-placed phone calls will usually identify sources of published information. For example, you may begin by consulting a librarian at your local reference library. The librarian can perform a computer search of the academic literature and of reports from commercial houses such as Predicasts, SRI, and Frost & Sullivan, and various commercial data banks. Purchasing subscriptions to such reports can be costly, but the value of the information may make it a justifiable expenditure.

Next, consult the relevant government department. You may find an industry expert who, if you're lucky, will have useful information at his or her fingertips. For example, in a preliminary study of the market for printed circuit boards, the researcher was fortunate enough to track down a government official who had just conducted a study on this industry. Although his findings were classified as confidential, a half-hour probing phone call enabled the researcher to obtain rough estimates of market size and growth (broken down by product type) and to identify the major competitors and their approximate market shares.

Finally, industry associations frequently maintain data banks containing survey results. In addition, the associations are helpful in identifying possible sources of information and key experts or leading firms in the industry.

Industry Experts

Usually, a handful of people possess most of the information you're looking for. For example, in the sound-absorbing brick case cited in chapter 3, acoustical consultants proved to be an invaluable source of information on the market and the product's chances for success. A number of phone calls and several quick visits to these experts yielded in one week information that might have taken a month or two to obtain otherwise. Similarly, in the preliminary market assessment for a toxic-waste-treatment facility, experts in several companies that had toxic waste problems were willing to share the information and research data that they had amassed.

Internal Sources

Surprisingly, market and product information is frequently available somewhere in your own company. Contact your sales force, your distributors and dealers, and your R & D group. Chances are someone will have had relevant experience, or will know someone qualified to speak knowledgably on the topic.

Potential Customers

Although it is premature to undertake a full-fledged market-research study at this early stage in the project, a handful of potential users can be surveyed about the market for the proposed product. A major packaged-goods firm regularly conducts informal "concept" tests of proposed new products at this stage of the game plan.

A simple questionnaire, together with a brief product description, is mailed to 100 customers, and the customers' responses to the concept are gauged. This same firm uses focus groups of potential users to judge the soundness of product concepts. (Note that in the packaged-goods business, market size is reasonably well documented; market acceptance or expected market share is the major concern in these preliminary assessments.) In the sound-absorbing brick example, an MBA student with an engineering background telephoned 50 potential users. The 20-minute interviews yielded data on market size, growth, pricing, and customer preferences. The cost was about $3,000, and the study took just over a week.

These examples show that a surprising amount of market information can be gathered quickly and relatively inexpensively. Much of the information is already in the public domain. The investigation work is not very scientific or very elegant — in fact, it's akin to detective work. Several phone calls are made; a handful of leads and names are obtained and followed up with more calls, and so on. Often this type of sleuth work can be handled by a persevering junior employee, properly directed. If you're short-staffed, the business school of your local university or college may be able to give you names of graduate students willing to work on an hourly or part-time basis. Sales trainees and recent retirees also are able to conduct this type of preliminary study.

Preliminary Technical Assessment

A preliminary technical assessment typically is an in-house endeavor. Several steps are involved:

- preparing rough specifications for the product;
- determining whether it is technically possible to develop and produce the product; and
- estimating cost and resource requirements for development and production.

While the results of the assessment will necessarily be speculative at this stage, a group of knowledgeable technical and production people usually can provide fairly sound opinions on these critical points, or at least identify areas of ignorance. An outside expert or technical consultant may be brought into the group as an additional source of information.

Preliminary Evaluation

The preliminary evaluation is the second GO/KILL decision point in the game plan, and follows the preliminary market and technical assessments (see Exhibit 6.1). The decision that must be made at this point is whether to proceed to the next stage — Stage 3 — where more extensive and expensive market and technical studies will be undertaken. The decision to GO at this point is still not a firm "GO into product development." But the information available at this preliminary evaluation is likely to be much better than that available at the screening stage, since informal market and technical assessments have now been completed.

Depending on the "goodness" of information available, the preliminary evaluation may start to rely on a financial assessment. If the market and technical assessments have yielded reasonable estimates of market size, possible market share or acceptance, selling prices, production costs, and expected investment in development and production, those data should be incorporated now.

Key Evaluation Questions

For many projects at this preliminary evaluation stage, the qualitative factors remain the overriding concern. The checklist or scoring model of "should criteria" can be used again. The key questions in this evaluation are as follows:

- Is the market attractive? Armed with more and better market information than at the screening stage, you should now be able to answer this question with greater confidence.
- Does the product have a differential advantage? Does it offer unique benefits to customers? Your technical and market assessments should answer these questions.
- Does the product fit the company — technically, from a production standpoint, and in terms of marketing? Again, your knowledge of what's involved in the development, production, and marketing of the product gained from the assessments sheds light on this question.

Some firms repeat the screening analysis using the scoring model ("should criteria" only) at this point. In particular, management looks for changes in the findings of the first and second screens: is the project looking better or worse than it did when it was first evaluated?

If the result of this preliminary evaluation is a GO decision, the project moves to stage 3. The identification of key questions, critical factors, and areas of uncertainty or disagreement are of the greatest importance. Their identification leads logically to the design of the market research and technical investigation studies of the next stage in the project.

Suggestion: Before proceeding with expensive technical and market studies, be sure to build the relatively inexpensive preliminary-assessment stage into your game plan. For example, assign a subordinate to do a market study in five to 10 days for under $3,000. When the market study is complete, convene a technical assessment session using inside people, augmented with a few outside experts.

Concept Definition

Concept definition is the final stage before product development begins — the last of the up-front activities, and perhaps the most critical one. At this point the product's target market, positioning, concept, and requirements must be finally defined to yield the protocol. Several actions are recommended (see Exhibit 6.1):

- *market research to identify the winning product concept:* a market study of buyers or users to define the requirements of the "ideal" product;
- *technical specification of the concept:* translating what the customers want or need into a technically feasible design or concept;
- *concept test:* taking the technically feasible concept back to the customers, and testing it to ensure that it is acceptable;
- *concept evaluation:* the final pre-development evaluation in which the protocol is defined and agreed to, and in which a GO decision takes the project into the product-development stage.

The Concept-Identification Market Study

The concept-identification market study is a key factor in arriving at a winning product concept and design. It is a prospecting study, seeking insights rather than soliciting feedback. The objective is to determine the "ideal product" in the eyes of the customer — the unique, superior product that delivers real benefits. The best way to sharpen the definition of this ideal is to consult the customer himself.

Before becoming too involved in the technical details of product design, try to ensure that certain market information is in hand:

- What product is the customer using now? Why? If your product is new to the world, what other product or products will it replace? This information gives an insight into who is buying what, how the customer perceives the product she is using at present, and what the customer's choice criteria are.
- What is the customer's level of satisfaction with the product she is now using? Why? This information shows how strong competitive brands or substitute products are, and their strengths and weaknesses. The identification of those strengths and weaknesses is often the clue to a superior design.
- What is the customer's level and order of preference for alternative competitive makes or brands? Why? This information identifies the deciding factors that cause the customer to choose brand X

versus brand Y; this understanding of choice criteria is critical to product design.

- How does the customer rate various brands on each of the choice criteria? Why? This information can be used to determine the positioning of competitive products in the market and the reasons for that positioning.
- What are the good and bad points of the competing products? What specific complaints or suggestions for improvement does the customer have? Again, this is vital information for the design team — knowing what competitors have done wrong, so that their mistakes can be avoided.
- What is the customer's ideal product according to each of the choice criteria? What features and attributes should be built in? This information helps to define the ideal product.

These key pieces of information are pivotal to the design of a winning product; they are the *information objectives* of a concept-identification market study.

The information objectives set out above call for a carefully designed market-research study. The example that follows illustrates how to carry out such a study. This example has been chosen to show that market studies are not limited to relatively simple consumer packaged goods, but also apply to industrial or technical products. Similar methods are used in concept-identification studies regardless of the product involved.

Step 1. Defining the Information Objectives

A manufacturer of heavy-duty trucks believed that there was a market for a new vehicle. A project team was assembled, and began to define exactly what information they wanted the market study to yield. Key questions focused largely on the design of the new vehicle. What should the product concept be? How should the vehicle be positioned in the market? What features and attributes should be built into the design?

Remember that information has value only to the extent that it improves a decision. The first step, then, is to define the decisions to be made. Then determine what information will contribute to the intelligent making of that decision. The truck manufacturer's project team identified many of the information objectives set out above. The team prepared a written list of objectives. (If your objectives are so vague that they can't be written down, you're headed for trouble before your study even begins!)

Step 2. Preliminaries

The project team was uncertain about what questions to ask customers, how they should be worded, and whether or not the customers would be able to respond to some of the questions. They also wanted to see how competitive brands were rated and viewed by users on criteria considered important to users. The team was not sure whether it had a complete list of these important choice criteria.

Discussions were held and focus groups convened to guide the design of the study. Meetings with the sales force and sales management personnel helped to answer some of the questions. Dealers' input also was sought. Finally, small focus groups made up of customers proved useful in completing the list of choice criteria.

Step 3. Research Design

The research method must be decided upon, the sample of respondents chosen, and the research instrument or questionnaire designed and tested. Survey research is the most popular method, and likely the most appropriate for concept identification (as opposed to experiment or observation). The project team chose personal interviews with fleet managers as the desirable method, in spite of its high cost. They also considered, but rejected, the idea of a telephone survey and a mail survey. (Appendix C highlights the pros and cons of the three survey approaches — telephone, mail, and personal interview — and gives suggestions for the design of each type of study.)

Following the focus-group sessions (used to identify key issues and the buyer's choice criteria), the appropriate questionnaire was developed and tested on a limited number of respondents.

The questions followed closely the list of information objectives that management had already developed. Exhibit 6.2 shows questions designed to determine how the various competitors rate overall in the customer's eyes, how each make rates on specific criteria, and why. Exhibit 6.3 focuses on the importance of these attributes or criteria to the buyer, what attributes he would like to see in his ideal product, and why each attribute is or isn't important. These two sets of questions often are key determinants of a new product's concept and its requirements. How the data obtained from these questions can be used in product specification is revealed later in this chapter.

In designing the study, the population of respondents must be defined: Whom do you want to gather information about? In the example, the population of interest was defined as "fleet managers of truck fleets of five or more heavy-duty class 8 vehicles." Defining the population, on the surface a simple task, can be problematic. For

Exhibit 6.2. Sample Questions to Gain Customers' Perceptions of Competitors' Products

---

PERSONAL INTERVIEW QUESTIONS:

(Note: this section of the questionnaire is preceded by a set of less threatening and easier-to-answer questions.)

1. Which makes of class 8 trucks are you aware of?

2. Which ones are you familiar with?

3a. If you were to purchase a new class 8 truck today, which make would it be? What would be your first choice?

3b. Why this make?

3c. Your second choice? (... and so on)

4. How do you rate the various makes? Let's take the makes you are familiar with... (spell these out). How do these makes compare or rate on the basis of a variety of characteristics? Let's start with suspension choice.

    You can answer by using this zero-to-10 scale. Score the make a 10 if it's excellent; score it a zero if it's very poor. (Show respondent the scale.)

    Very poor   0 1 2 3 4 5 6 7 8 9 10   Excellent

    Let's begin. How would you rate (make) in terms of suspension choice? (Repeat for all familiar makes, and for list of characteristics below; record answers in table below.)

Characteristics:	GMC	Ford	Navistar	Mack	Kenworth	White	Freightliner
Suspension choice	___	___	___	___	___	___	___
Frame strength	___	___	___	___	___	___	___
Outside appearance	___	___	___	___	___	___	___
Interior appointments	___	___	___	___	___	___	___
Weight savings	___	___	___	___	___	___	___
Etc...							

5. (For those items in question 4 above that featured very high or very low answers.) Tell me, why do you answer so high (or so low)? What made you respond so strongly?

---

example, in defining the population for a study to design a new lawn-and-garden tractor, any one of the following definitions might be used:

- homeowners;
- homeowners with more than one acre of land;
- current owners of lawn-and-garden tractors;

Exhibit 6.3.   Sample Questions to Measure the Importance of Attributes and the Ingredients of the Ideal Product

---

1. Here's a list of characteristics or features of a class 8 truck — features that may or may not be important to you in deciding which make of truck to buy. I'd like you to consider how important each is in your purchase decision.

   You can answer by using this zero-to-10 scale. Here, 10 means it's of critical importance and zero means not important at all. (Show respondent the scale.)

   Not important   0   1   2   3   4   5   6   7   8   9   10   Critical
   at all   importance

   Let's begin with suspension choice. How important is having... etc.

Characteristics	Importance Rating	Reasons
Suspension choice	_____	_____
Frame strength	_____	_____
Outside appearance	_____	_____
Interior appointments	_____	_____
Weight savings	_____	_____
Etc...		

2. Why did you rate (characteristic) the way you did? (record answer above)

3. (For each characteristic above) What improvements would you like to see in terms of (suspension choice)?

4. Imagine that you were a member of a design team charged with coming up with a class 8 truck designed for the user and buyer. What three recommendations would you make for the new design?

- current owners of all power lawn-mowing equipment, including power lawn mowers, riding lawn mowers, lawn tractors, and larger lawn-and-garden tractors; or
- owners of that company's brand of lawn-and-garden tractors.

After the population is defined, a method of sampling must be devised. In the example, a listing of fleets was available, so it was relatively simple to select at random respondents to be interviewed. Geographic locations (in this case cities) were selected first, and respondents from each city were chosen. A stratified sampling method was used — a refinement of random sampling whereby the sample is biased toward certain strata or subgroups. In this case, the sample was stratified by fleet size and biased toward the larger fleets to take account of their greater importance to the total truck market.

Step 4. Implementation

Approximately 100 personal interviews were conducted in 10 cities in the truck study. Five interviewers were used, and each was thoroughly trained prior to field work. Management sat in on the first few interviews and stayed in touch with the field force throughout the study.

Step 5. Analyzing the Results

As shown in Exhibits 6.2 and 6.3, many of the questions were scaled questions requiring a numerical answer. (There also were many open-ended questions requiring a verbal and more lengthy response; open-ended questions will often lead to the richest information.) The number of questionnaires and the numerical nature of many responses made it economical to use a computer for analysis. (In the case of some open-ended questions, response categories can be developed by reviewing the first several dozen questionnaires, and those responses can then be coded.)

The questions shown in Exhibit 6.2 sought information on how the respondent viewed the competitive makes of trucks — overall and on a list of choice criteria thought to be important in the buyer's choice of make. The following types of analysis and tabulations were undertaken.

Relative preferences. Relative preferences were obtained for each make: which competitor was the most popular, second most popular, etc. Preferences were broken down by type of buyer: large fleet versus small, east-coast versus west-coast, public carrier versus private fleet, and so on. Knowing which segments prefer which makes of trucks for which reasons is crucial to the marketing effort.

Ratings of brands on various criteria. Ratings of trucks were obtained on each of the choice criteria. Which make was perceived to have the highest resale value? The lowest? Why? Which make has the best fuel efficiency? The worst? Why? Exhibit 6.4 shows how each make scored on each criterion. The probing questions — "Why do you see a Kenworth or a Mack as being the most reliable vehicle?" — prove invaluable in identifying what it takes to score high on each criterion. (The ratings shown in Exhibit 6.4 were broken down by various categories of respondents — large versus small fleets, various geographic regions, and types of fleets — so as to identify differences in perceptions between buyer segments.)

Exhibit 6.4.   Profiles of Various Truck Makes on Choice Criteria

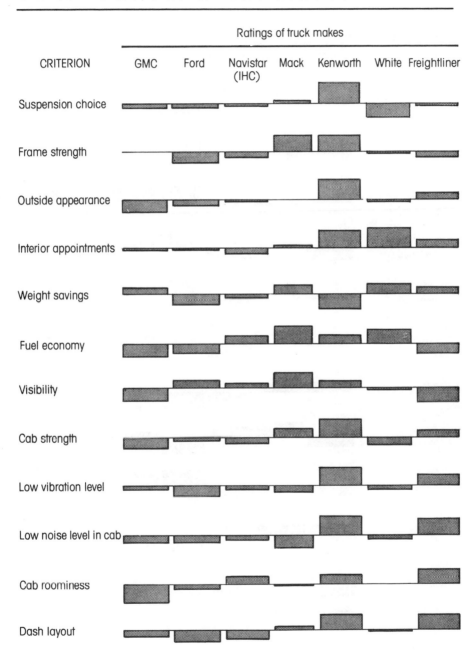

(Continued)

Exhibit 6.4 (continued)

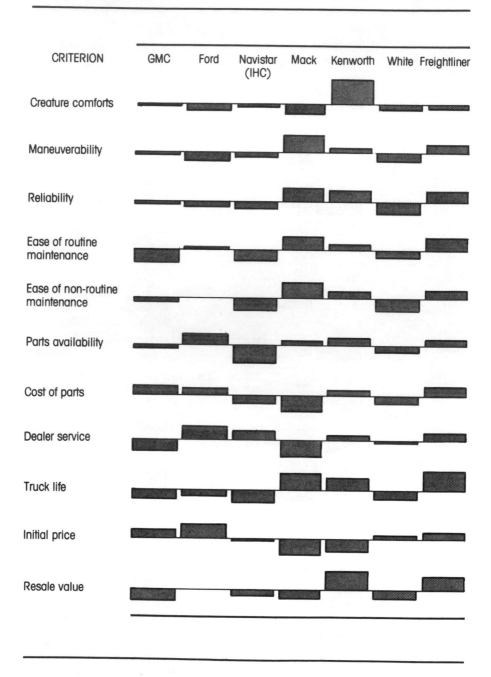

CRITERION	GMC	Ford	Navistar (IHC)	Mack	Kenworth	White	Freightliner
Creature comforts							
Maneuverability							
Reliability							
Ease of routine maintenance							
Ease of non-routine maintenance							
Parts availability							
Cost of parts							
Dealer service							
Truck life							
Initial price							
Resale value							

Underlying evaluative factors. Rarely does a buyer of any product actually evaluate brands on a long list of formal criteria in order to make his or her purchasing decision. Rather, customers tend to simplify the world and think in terms of "evaluative dimensions." For example, one familiar evaluative dimension in the purchase of many products is "economy," which includes the criteria of low purchase price, low operating and servicing costs, high resale value, and so on. To identify the evaluative dimensions for the truck study, it must be determined which criteria are closely related to each other. For example, three criteria — ease of routine maintenance, truck life, and reliability — are closely connected (based on correlation coefficients). There is likely one underlying evaluative dimension that captures or explains the presence of these three criteria.

A familiar statistical technique called factor analysis was used to look for underlying dimensions. (The input variables in the factor analysis are the choice criteria, and specifically the ratings of all makes on those criteria.) Six underlying dimensions were found in respondents' perceptions of makes of trucks. Exhibit 6.5 shows four of the six dimensions in pictorial form, and indicates how the original choice criteria or attributes are correlated with these dimensions. For example, such attributes as ease of routine maintenance, few breakdowns, length of truck life, and ease of nonroutine maintenance are all related to one another and together constitute the global dimension called "dependability." A second evaluative dimension was "cab design," which consists of vibration and noise levels in the cab, cab roominess, and cab strength. The maps shown in Exhibit 6.5 are the first step toward an understanding of the reasons for the market positions of competing makes.

Competitive positions. A positioning map shows the buyers' perceptions of how various brands or makes are positioned relative to one another. A knowledge of the evaluative dimensions, the attributes or criteria that comprise each dimension, and how competing products are rated on these criteria permits the construction of such a positioning map. Exhibit 6.6 shows that Mack was perceived as the best make on the dependability dimension, with GMC faring poorly. Kenworth featured the best cab design, with GMC and Ford faring poorly. The second map shows Kenworth exceptionally strong on the aesthetics dimension, with GMC again doing less well. Some obvious questions arise. Why is Kenworth so strong on some dimensions? Why is Mack strong? What's wrong with GMC's truck design? Answers to these questions are obtained by returning to the original ratings (Exhibit 6.4) and the respondents' reasons for the ratings.

Exhibit 6.5. Evaluative Dimensions in the Truck Market

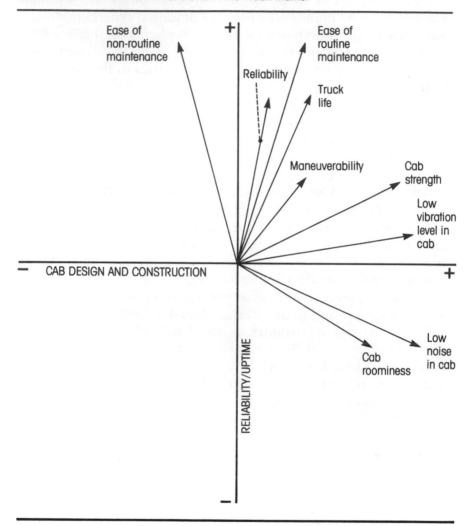

These answers provide guidance to the product-design team about what elements should be built into a winning new truck design.

The questions shown in Exhibit 6.3 focused on characteristics buyers considered important in a truck. These sample analyses help to define the ideal product.

Prioritization of choice criteria. Respondents were asked to indicate the relative importance of the choice criteria in the selection of a vehicle (see Exhibit 6.3). Those scores could be simply tabulated, as shown in Exhibit 6.7, to yield a profile of factors considered impor-

Exhibit 6.5 (continued)

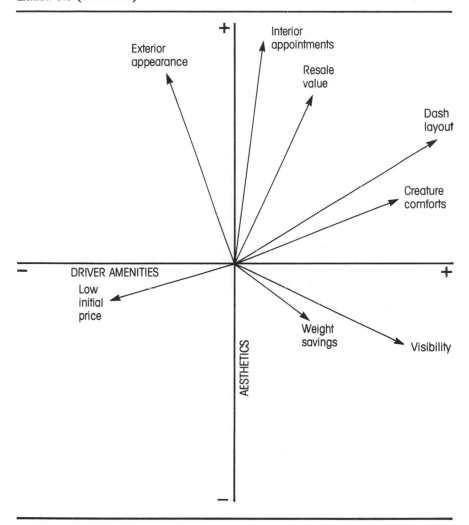

tant in a new truck design. Perhaps the greatest insights come from the verbal and open-ended responses. When respondents said that something was particularly important to them, they were asked why. The reasons they gave added depth to the numerical analysis. The importance ratings were also broken down by market segment; Exhibit 6.7 was re-created for large fleets versus small, by geographic region, and by fleet type to see how perceptions of what is important change by buyer type.

The ideal product. By combining the market-positioning maps

Exhibit 6.6.   Positions of Various Truck Makes

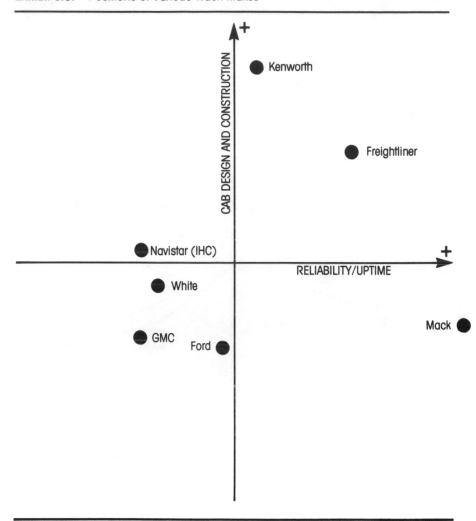

shown in Exhibit 6.6 and the knowledge of buyers' choice criteria, the "ideal product" can be identified in terms of key evaluative dimensions. Of course, no two buyers will have the same "ideal." (In the example, about 100 "ideal" products were defined.) By visually clustering or grouping these ideals as shown in Exhibit 6.8, we end up with several different ideal trucks, each one desired by a different type of truck buyer. (In the example, a statistical method called cluster analysis was used to group the ideals.) In effect, different groups of buyers in the market, with different notions of the ideal product, were

Exhibit 6.6 (continued)

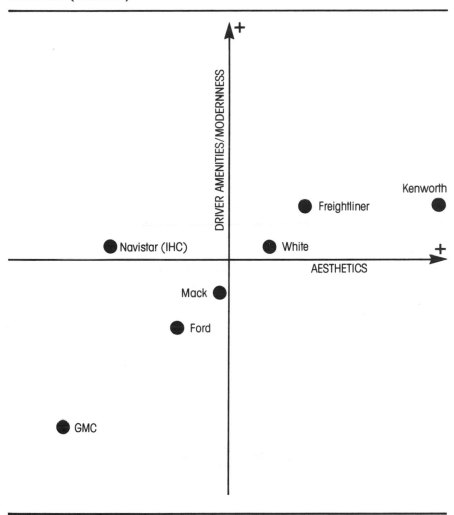

identified. Those groups are market segments, or, more precisely, benefit segments. The task is to decide which benefit segment should be targeted with the new product. That decision is made by reviewing competitive positions (Exhibit 6.6), their strengths and weaknesses (Exhibit 6.4), the size and composition of alternative target markets (Exhibit 6.8), and the opportunity to gain a differential advantage in the market segments. Having selected a target market and a positioning strategy, we now return to identifying the attributes that will enable us to build that position. The data in Exhibit 6.5, together

Exhibit 6.7.   Mean Importance Ratings of Choice Criteria

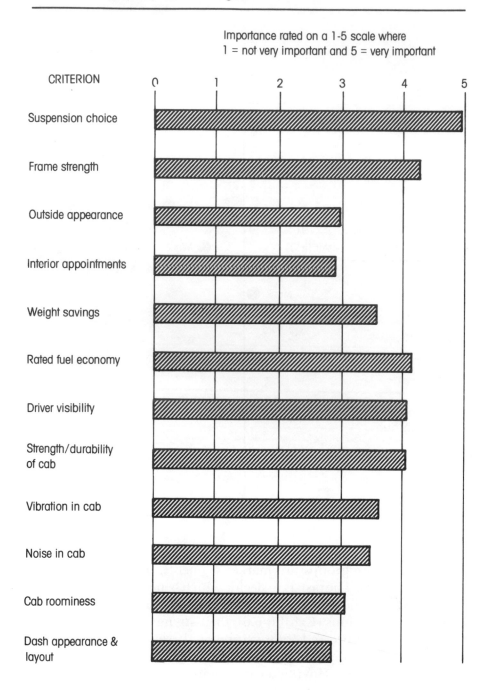

Importance rated on a 1-5 scale where
1 = not very important and 5 = very important

| CRITERION | 0 | 1 | 2 | 3 | 4 | 5 |

Suspension choice

Frame strength

Outside appearance

Interior appointments

Weight savings

Rated fuel economy

Driver visibility

Strength/durability of cab

Vibration in cab

Noise in cab

Cab roominess

Dash appearance & layout

Exhibit 6.7 (continued)

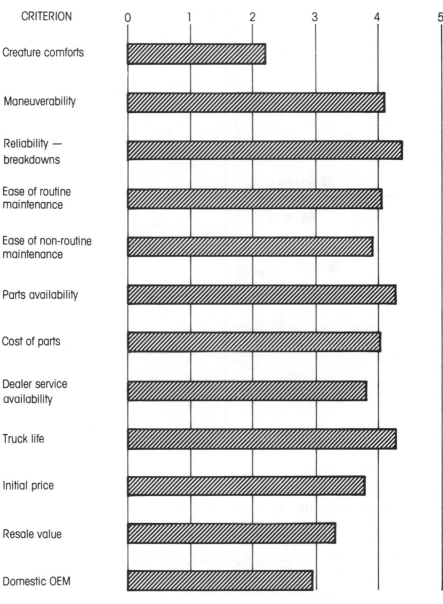

Importance rated on a 1-to-5 scale where
1 = not very important and 5 = very important

Exhibit 6.8. Benefit Segments and Possible New-Product Positions in the Truck Market

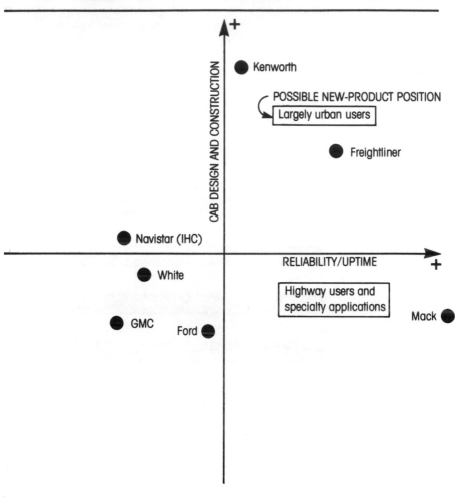

with the answers to the open-ended questions, provide the design team with enough information to put together a fairly detailed concept statement and set of design requirements.

Alternative analyses. Some types of questions and analyses were not a part of the truck study, but are often included in a concept-identification market study. For example:

- In addition to asking how important each attribute was to the buyer, we could have asked the buyer to indicate the degree or

Exhibit 6.8 (continued)

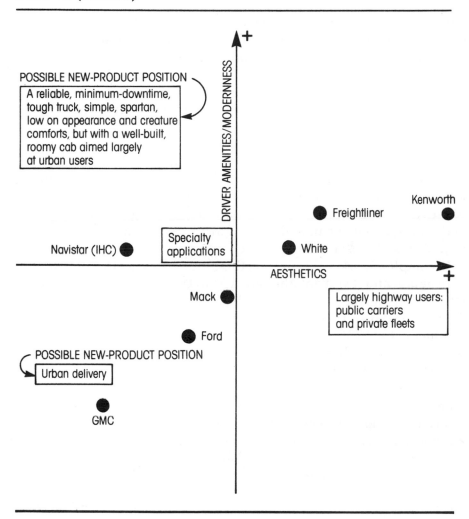

amount of each attribute he would want to see in his preferred truck. In short, the buyer is asked to design his ideal product.

- If the product was a simpler product, such as a packaged good, if there were many brands on the market, and if buyers were familiar with a good number of brands, the ideal product could have been inferred by relating stated brand preferences to each brand's position on the evaluative map, and then using multiple regression, for example, to determine which dimension was the most important.[2]

Concept-Identification Market Studies

The sample questions shown in Exhibits 6.2 and 6.3 are only a few of the many possible questions that can be asked of potential customers. Answers to open-ended questions provide the richest data in terms of design specifics and customer needs and preferences. These answers are frequently tape-recorded during the interview and played back to give the design team firsthand information about what the customer is thinking.

Some of the statistical analyses and the resulting positioning maps may seem complicated, particularly if your firm has had little experience in conducting these types of market research studies. But don't let the sophistication scare you off. Such research is invaluable in identifying the target market, the positioning strategy, the new product's concept statement, and the product's desired attributes — the ingredients of a protocol statement.

If complex market research methods are likely to frighten your colleagues, and if you and they are new to such market studies, start with a simpler study, one that involves relatively easy-to-understand data analysis. Convene focus groups to identify the key issues in the product's design from a customer's standpoint. Then, conduct a larger sample study by mail, telephone, or personal interview that does the following things:

- determines what the buyer is using now, why he bought it, the buyer's brand preferences, and the reasons for those preferences;
- identifies his likes and dislikes in current products on the market, especially the one he uses, and solicits suggestions for improvement in product design;
- obtains ratings of current products on a list of choice criteria to yield a competitive profile table (similar to that shown in Exhibit 6.4) and the reasons underlying these ratings (particularly in the case of extreme positive or negative ratings);
- obtains an indication of the relative importance of the choice criteria to yield a profile table similar to that shown in Exhibit 6.7, and the reasons certain items are important;
- obtains some descriptive information about the buyer that will help to segment the market: age, sex, income, place of residence, frequency of purchase, amount purchased, etc., in the case of consumers; and firm size, location, business, frequency of purchase, amount purchased, etc., in the case of industrial clients.

Suggestion: Does your firm build a concept-identification market study into your new-product projects — one whose objective is to determine the winning product design or concept? If not, consider doing so on your next project; it could be the key to your success. First, identify the key decisions that you face (many of these will be product-design and product-strategy decisions). Next, write down your information objectives — what market information would help you make those decisions. Follow this by the research design, enlisting the services of a professional if necessary. If this is your first attempt at such a study, you may wish to avoid the sophisticated version outlined in the trucking firm example and simply use the five questions listed above. If you're more experienced, try advancing to a positioning and benefit-segmentation study similar to the one in the example.

Concept Development

The type of market study outlined above provides useful guidelines as to what should be built into a new product's design. Properly designed and executed, your market research should indicate what the "winning product" is from the customer's perspective — the unique and superior product that delivers real benefits to the customer.

The customer "wish list" must now be translated into something that is technically and economically feasible. This is where the marketers (those who have direct contact with the customer) and the designers (those who will design and develop the product) must marry their thoughts to arrive at a proposed product concept. In short, the designers must find a means of satisfying expressed customer needs and preferences. This is a creative process, and one which is difficult to formalize. Consider the following illustrations.[3]

A consumer-goods firm wishes to enter the lucrative window-cleaner market with a new product. The product idea was born in the marketing group, largely because of the magnitude of the market and the dominance of a single competitor's brand. The idea for the product is fuzzy — the marketing people envisage a stronger, heavy-duty cleaner that "works better" and "smells nicer." This is the extent of the creative thinking that has been done. The lab people have been experimenting with various formulations, and preliminary market assessment reveals some buyer interest and a potentially lucrative market. The ingredients of a winning product design are still missing, however.

A concept-identification market study is commissioned, in the form of personal interviews with householders at their homes, in the hope of identifying the missing ingredient. The market study reveals that ease of use, cleaning efficacy, freedom from streaks, pleasant smell and color, and price are major criteria in the purchase of a window-cleaning product. Further questioning shows that the leading brand scores exceptionally well on all of the criteria except price, and private brands score well on price. It is decided that the new entry will not be a low-priced brand, and that it will have to compete on the basis of product advantage. Against a formidable competitor, there appears little the new entry can offer. Are we heading for another "me too" entry doomed to mediocre sales results? Not necessarily.

The same market study reveals some of the leading brand's weaknesses (the weaknesses are identified largely in the open-ended and probing questions, and in some cases by watching householders clean their windows). The leading brand is too wet — it takes several rolls of paper towels to clean and dry all the windows in a house, more than the cost of the cleaner. The cleaner dries poorly, particularly outdoors and in cool weather. It cleans well most of the time, but doesn't remove the dirt and grime on outside windows, particularly the dirt that collects behind screens. Three hands are needed to clean a window: one to hold the roll of towels, one to hold the cleaner, and one to wipe the window. Finally, although the question was not formally asked in the questionnaire, the study found that most people had a strong aversion to cleaning windows at all — no big surprise!

The designers, chemists, and engineers meet with the marketers to discuss the results of the study and to engage in concept development. Because participants were armed with the market information, there was some direction to the session: everyone was looking for solutions to real problems. Some ideas begin to emerge, and the following products were suggested:

- a spray cleaner that goes on relatively dry, and turns into a light, dry powder that can be wiped off easily without soaking the towel;
- a disposable nonwoven wiper cloth impregnated with a window-cleaning agent;
- a heavier-duty cleaner or an adjustable-strength cleaner (to be mixed with water in varying proportions) for light or heavy jobs;
- a bottle design that enables the container to be held in the same hand that's doing the wiping;
- a spray or film that repels dirt, reducing the need for frequent window cleaning.

These are the types of concept statements that should emerge from a concept-development meeting: technically feasible solutions to the customer's problems. Of course, all of the ideas will require further thought and technical work to refine and prove their feasibility. The end result should be a statement of a technically and economically feasible product design that responds to customer problems and needs and is demonstrably better than competitors' products. When that statement is complete, it's time to test the concept — to verify that the customer's needs are met by a poduct design that is not only acceptable to him or her, but is strongly preferred.

The Concept-Test Market Study

Will the new product be a winner? Before pushing ahead with product development, you must be certain that the product will meet customer needs and wants better than the competitors' products. Remember that your product is the new entry into the market, and it has to give the customer a reason to switch.

A problem faced by many firms is one of translation. A thorough market study is undertaken that identifies customer needs and wants. The technical team then translates those into a tentative product design — but something goes wrong in the translation. The final product isn't quite what the customer wants, or it lacks that special something that differentiates it from what the customer is already buying; it just doesn't push the customer's "hot button."

Prospecting Versus Testing

It makes sense to build in a concept-test stage before proceeding to product development (see Exhibit 6.1). The concept-identification market study (for example, the truck study) is a *prospecting* one: no product was available to show the customer, but hints, clues, and evidence were obtained from the customer about what should be built into a winning product design.

Once the technical concept-development work has yielded a feasible design, a model, a drawing, or a written list of features can be shown to the customer and her response gauged: "Given what you've told us, this is the proposed product; now what do you think of it? Would you buy it?"

Note that at this stage you still don't have a fully developed product. The purpose of the concept stage is to see if you're heading in the right direction. By this time you should have at minimum a written description of the product and its features, benefits, and likely price. In addition, you may have something concrete to show the customer — line drawings, artist's renderings, a model, a slide show or a rough prototype of one or more product concepts.

Designing a Concept Test

The design of a concept test is similar to that of the first market study. At minimum, you might use a handful of focus groups to gauge reactions to the proposed product. While such focus groups give useful feedback on the product, remember that the limited sample size, the fact that group members are often self-selected, and the nature of group dynamics means that the group's views may not be totally representative of your target market. Often a survey by mail, telephone, or personal interview is the more desirable route to take.

The major difference between the concept-test market study and the concept-identification market study is the type of information sought. In a concept test, the information objectives typically include:

- a measure of the customer's interest in the proposed product or products and a determination of why the interest level is high or low;
- a measure of the customer's liking for the proposed product or products;
- a measure of the customer's preference for the proposed product or products relative to competitive brands or the product the customer now uses, and the reasons for this preference;
- an indication of the customer's intent to purchase the proposed product at a specified price.

Exhibit 6.9 shows a typical questionnaire format. It is good practice to design a standard format and to use that format consistently from product to product. In this way you will develop a history of data and establish benchmarks for comparison. For example, when 30 per cent of those surveyed reply in the "definitely would buy" category, what does this mean — is it good or bad, and what market share might this response translate into?

Armed with the results of the concept identification truck market study referred to earlier, the engineering and design team were able to develop several products that were technically and economically feasible and responded to the customer's desires. Eight product concepts were developed, largely variations on a single theme.

Exhibit 6.9.   Typical Questions in a Concept Test

Face-to-face interview:

(The respondent is asked to look over the proposed product concept or concepts… a written description, or a sketch or drawing, etc.)

1a. First, what's your reaction to the proposed product? You can answer using this zero-to-10 scale, where zero means very negative and 10 means very positive. (Show him/her the scale.)

Very negative   0   1   2   3   4   5   6   7   8   9   10   Very positive

1b. Why so positive (or negative)? _____

2a. How interested are you in the concept? (Show response category scale.)

_____	_____	_____	_____	_____
not interested at all	not too interested	somewhat interested	quite interested	very interested

2b. Why did you answer the way you did? _____

3a. To what extent you like the proposed product? Please answer on this zero to 10 scale, where 10 means "like very much" and zero means "don't like at all."

Don't like it   0   1   2   3   4   5   6   7   8   9   10   Like it very
at all   much

3b. Why did you like/not like it? _____

4a. What is the likelihood that you would buy this product at a price of $XX?

_____	_____	_____	_____	_____
definitely not	probably not	maybe	probably yes	definitely yes

4b. Why or why not? _____

5a. What do you see as the product's main strengths? _____

5b. Its main weaknesses? _____

5c. Would you like to see anything changed? What are your suggestions? _____

A concept test was then undertaken. An industrial design firm was retained to translate the product concepts, now in written form, into visuals. Eight cardboard displays, each about 20 to 10 inches, were prepared, one for each concept, which demonstrated how the truck would look, inside and out, and listed the features and specs of each. A price was suggested for each.

Seventy-five potential users in the defined target market were asked to review each concept, and give their gut reactions to them. (These taped interviews yielded some rather colorful and direct responses.) Second, the respondents were asked the four questions shown in Exhibit 6.9 for each concept. Then they were asked to sort the display cards into two piles — those they would buy and those they wouldn't — and to indicate why as they sorted. Finally, they were asked to go through the "would buy" pile, and rate the choices: which product would they buy first, second, third, etc., and why.

The results not only indicated which concept was preferred and by whom, but also revealed the reasons. The strength of acceptance and the features or benefits the customers were responding to were also obtained.

Using the Results of the Concept Test

Use the results of a concept test with caution. They merely provide an indication of likely product acceptance — there are no guarantees. Nor should the results be used blindly. For most new products, concept tests are likely to overstate the market acceptance. For example, the statements "30 per cent definitely would buy" is not likely to translate into a market share of 30 per cent, for several reasons. First, respondents tend to have a positive response bias. There are many reasons for this: the so-called Hawthorne effect, whereby people under observation tend to respond more positively or enthusiastically than those not being studied; the desire to give socially acceptable or pleasing answers to the interviewer; and the fact that it's easy to say yes when no money or commitment is involved. Second, although the respondent may say that he'll buy your product, in the case of a frequently purchased product he may continue to buy the competitor's as well. If he buys both equally, the "30 per cent definitely would buy" response actually translates to a 15 per cent market share. Third, not all potential buyers will be exposed or exposed sufficiently to your new product. (In a survey, all respondents have a chance to see the proposed product.) Your advertising, promotion, sales force, and sampling program may reach less than half of the total target market. The "intent to purchase" figure must be cut down by a factor that reflects your market exposure on launch. Fifth, for more innovative products, the concept test may understate the product's acceptance. For example, in the case of a new milk-packaging system involving plastic bags instead of bottles or jugs, the concept was so alien to potential users that they replied negatively. Once the concept had been explained and demonstrated so that the user understood the product and its benefits, reaction was much more positive. In a similar vein, a concept test of a proposed office-of-the-future com-munications system involving CRTs, desktop computer keyboards, electronic mail, electronic messaging, and direct access to in-house and remote computer data banks gave negative results. The concept came across to the average users — office managers — as a "Buck Rogers" product: they simply couldn't relate to it.

With all its weaknesses, the concept test is still a useful means of gauging likely product acceptance. Perhaps even more important is the information obtained from the open-ended questions, "Why did you like it?" and "Why don't you like it better than Brand X?" The answers provide significant insights into what needs to be done to make the product as attractive to the buyer as possible. Further, the comments of potential customers provide insights into how the product might best be positioned and communicated to users — what "hot buttons" have been hit.

Suggestion: Even though you may have conducted a proficient con-cept identification market study and learned exactly what the ideal product is in the customer's eyes, and even though you may feel that you've done an honest job of translating this into a technical concept, it's always best to check it out. Undertake a concept test. Find out whether your proposed product really is "right" from the customer's perspective. Measure interest, liking, preference, and intent to pur-chase, and find out why the customer responds the way she does to your concept. With this information in hand, you will be in a better position to define your protocol and proceed to the product-development stage.

Concept Evaluation

Concept evaluation is the final GO/KILL decision point before mov-ing into full-scale product development (see Exhibit 6.1). The results from the concept-identification market study, the technical aspects of concept development, and the results of the concept-test market study are combined. If the decision is GO, these findings will be integrated into a protocol statement.

The GO/KILL Decision

The GO/KILL decision here is a critical one. Once you pass this point in the game plan, it becomes increasingly difficult and expensive to turn back. The GO/KILL decision must involve a combination of qualitative and financial considerations. The checklist of questions used in screening and preliminary assessment can be repeated. (Some firms simply use their scoring model here: the assumption is that by this point the evaluators are better informed and the screening tool is likely to give more accurately predictive results than at earlier stages in the project.)

Because the next stage, product development, involves major expenditures, a financial analysis of the project is warranted at the concept-evaluation stage. (Financial analysis is dealt with in chapter 7.) The protocol is the second decision (or set of decisions) to be made at this concept-evaluation stage. To recap, the protocol is an agreement between all parties that guides the product-development phase and charts the overall course for the remainder of the project. It specifies the target market, the positioning strategy, the product concept and benefits, and the product requirements.

On to Product Development

With the GO decision made and a protocol in place and agreed to, the up-front activities are complete, and product development can now begin in earnest.

Armed with the information contained in the protocol, the design team can go to work. They know who the customer is and the product that he wants, needs, and prefers. They understand the competitive products and what their good and bad features are. They know what benefits the new product must deliver to make it a winner, and they know how it must be differentiated from competitors' products in order to gain an advantage in the market. Contrast this situation to the information vacuum that the Gemini team faced when they began development work on the printer.

How well the development phase is carried out will depend in part on the skills, knowledge, resources, and management prowess of the development group. But at least you will have paved the road for them and given them a fighting chance for success. Note that even in the product-development stage, and in spite of a clearly defined protocol, questions and issues will arise that may require occasional cycling-back to earlier stages. For example, during the design and development phase, design questions — which is the preferred layout of an instrument panel? which package does the customer like best? — will be faced. It is possible, and in fact desirable, to involve customers by undertaking mini-concept tests in which customers are asked for their reactions to aspects of the product's design.

It's GO for product development! With a well-defined protocol in place, a reasonably skilled and managed development group, and some degree of luck, the result should be a finished prototype or sample product. You'll be well on your way to successful product development.

Testing the Product and the Strategy

The project is GO for development. The up-front homework has been done, and the project has been clearly defined in the protocol. As experienced project managers will attest, however, even in the most astutely defined project much can go wrong from this point on. The market may change partway through development, making the original estimates of market size and acceptance invalid. Competitors may introduce a similar product in the meantime, or users' preferences may shift as the product nears the final stages of development. Another common problem is that the final product may not receive the same enthusiastic reception from potential customers that the concept did. This apparent discrepancy may occur because the development team incorrectly translated the concept into a product, because technical problems were encountered that forced some design changes, or because errors of interpretation were made. Or, in the concept test, the customer may have been responding only to a concept — the tangible product or prototype did not, in the customer's perception, "match" the concept.

No Surprises

One can never be sure about the success of a product until it actually goes on sale in the marketplace. Thus, it becomes imperative to build into the game plan a number of checkpoints and tests to make sure that the project is still on target as we move through development and towards market launch. This is the topic of chapter 7: how to build in those checks and evaluations during and following product development. The name of the game is "no surprises!"

In-House Product Testing

Development work is underway. The product is taking shape. Depending upon the length of the development process and how that process is structured, product or component testing will likely play an important role, certainly at the completion of product development and possibly even during the development phase.

This type of product testing is familiar: the prototype is subjected to a set of rigorous engineering and laboratory tests to make sure that the product is functioning correctly. In the case of a lengthy development process, it makes sense to subdivide the process into discrete sections, and to build in product or component tests, where possible, at the completion of each development step.

Most firms perform this type of product testing as a routine part of the product-development process. Exhibit 7.1 depicts this testing as taking place after product development is completed. Recognize, however, that prototype or component tests can also be undertaken at the completion of individual steps during the product-development process.

Customer Testing

An often-forgotten facet of product testing is the customer test. An in-house product test only confirms that the product works properly under controlled or laboratory conditions. It says little about whether the product works under actual use conditions, and whether the customer finds it acceptable. Customers seem to have an innate ability to think of novel ways of finding product weaknesses, ways the engineering-testing group could never have imagined. The acid test of a product design is with the customer.

The point is this: not only must the product work right in the lab, it must also work right when the customer uses and abuses it. The product also must be acceptable to the customer (simply "working right" doesn't guarantee customer acceptance). Finally, the product must excite the customer: he or she must not only find it acceptable, but must actually *like* it better than what he or she is buying now. Customer testing should be an integral part of your game plan. At a minimum, customer tests should be performed when the prototype is complete; ideally, they should be an ongoing part of the entire product-development process.

Doing Customer Tests

How does one design a customer test? Here are some relatively simple tests you can build into your game plan.

Assume that you are partway through the development of a fairly complex product — for example, a new lawn-and-garden tractor aimed at homeowners. Key components — the new automatic transmission and the dashboard instrument panel — have already been designed, developed, and tested in-house. Both of these components are highly visible in the final product: how the transmission shifts, and how the dash looks. Here's what you can do to determine the degree of customer acceptance:

- Bring potential users (and your dealers) to the development site (or to a convenient location, such as a suburban hotel) to view and try out the key components. You might mount the transmission on an existing tractor and display a mock-up of the dash available. Let the customers look, touch, and try. Record their reactions and comments (on tape, if you can). Then interview them. Obtain basic background information (demographics and other segmentation data), and then measure interest, liking, preference and intent (using the question format shown Exhibit 6.9). Include probing questions, noting particular areas of likes and dislikes. If the customer has problems and voices complaints when he tries the product or component, note those as well.

- The same procedure can be used with focus groups of customers. Start with an introductory group session. Then move to the display area, so that the customers can touch and try. Finally, reconvene the group for a discussion of the merits and shortcomings of the tested components. The group session is more efficient than individual interviews (more inputs in a shorter time), and often leads to a more interesting discussion (the group members stimulate one another), but be careful of group dynamics: a single powerful member can sway the whole group to a positive or negative reaction to the component.

- When the number of customers is small, try setting up a "users' panel" — an ongoing group of potential customers that acts as a sounding board or team of advisers during the development process. Whenever designs, design decisions, or components need to be checked, convene the group to get its reaction.

Ongoing Customer Research

Don't be afraid to reach out to large numbers of customers to answer key design questions that arise during the development process. The original market research that was done prior to product development may not be enough to resolve all your design dilemmas. Technical problems may arise during the development phase that necessitate a significant product-design change. If the impact of the change is

Exhibit 7.1.   The Game Plan: From Development to Market

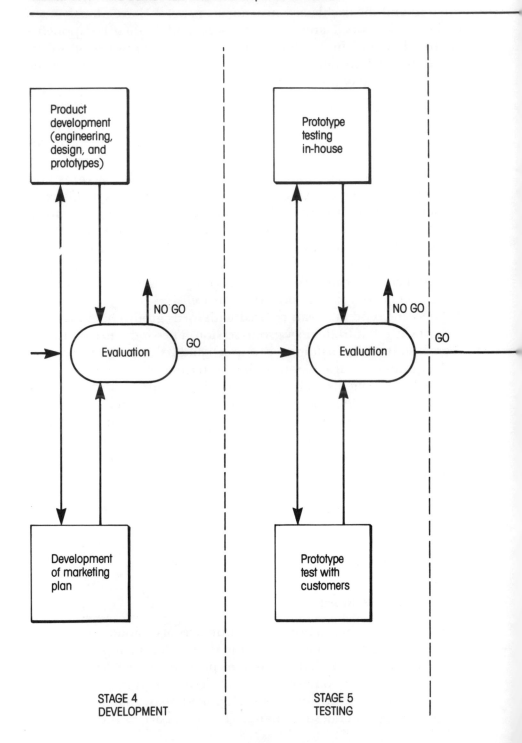

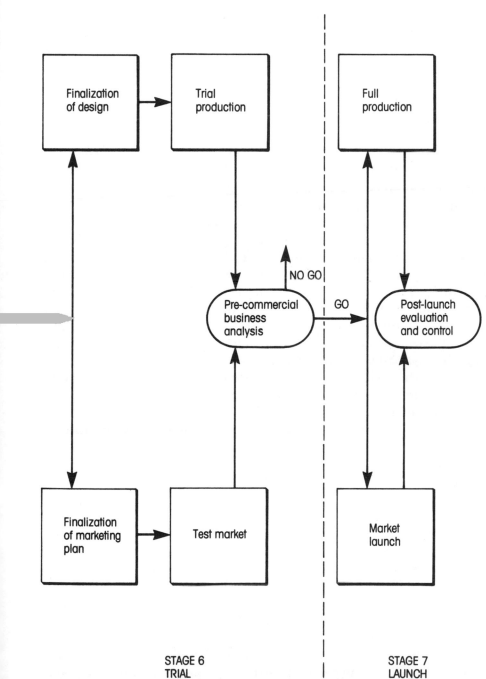

STAGE 6
TRIAL

STAGE 7
LAUNCH

Adapted with permission from R.G. Cooper, "A Process Model for Industrial New Product Development," *IEEE Trans. Engineering Management* EM30 (Feb. 1983): 2–11. Copyright ©1984 IEEE.

likely to be visible to the customer, check it out — don't assume. If market conditions are changing as a result of changes in customers' tastes and preferences or because a new competitive product hits the market, don't be an ostrich. There's no disgrace in admitting that the up-front market investigations didn't answer every possible question. It's a fact of life that conditions change and unforeseen technical problems occur. The Edsel was one of the most carefully market-researched automobiles. The research was all done prior to development. Unfortunately, customer tastes changed, and so did the economy. Moreover, key design decisions were rarely checked out with customers — the styling of the car, the electric pushbutton transmission located in the steering-wheel hub, even the name of the car. What would the outcome have been if Ford had reached out to its customers at every step of the design and development process?

Astute product developers recognize that additional market surveys may be required, even during the development process. For example, in the development of a milk-packaging system DuPont's development team hit a technical snag.[1] The original concept was for the plastic milk bag to have a tear-off tab for easy opening and resealing. The pre-development market research was done on the assumption that this design would be implemented.

During development, however, technical difficulties arose that made the tear-off tab almost impossible. Rather than merely assuming that a change in product design would be acceptable to the customer, the design team asked the marketing department to survey users to determine the importance of the tear-off tab. A hastily commissioned market survey revealed that the tear-off tab was desirable, but not essential, and product acceptance would not be significantly affected by its absence. The tab was removed from the design.

Testing After Development

For some products, the first time the customer can see and try the product is after the prototype or sample is manufactured. This is true in the case of relatively simple products, such as packaged goods. It may also be true for more complex products, where the product is such that individual components cannot be tested separately — the components must be combined before the customer can get a feel for the product. An example of the latter is an office telecommunications/information system, where the separate components, including software, desktop hardware, large switching devices, and communications lines, mean very little to a customer until they are operating together as a system.

By delaying customer testing until the end of development, the risks increase. Try to build in customer tests during development. For example, customer testing of an office information system would let users see and try a desktop unit, a simulated display or output, and so on. Remember that the objective is to test and check the product often — no surprises! Here are some typical tests you can undertake with the finished prototype or sample.

Preference Tests

A preference test, in which customers, either individually or as a group, are brought to a convenient site and exposed to the product, and their interest, liking, preference, and intent measured, does several things. First, it provides a more accurate reading of likely market acceptance than the pre-development concept test. In the concept test, the customers saw only a description of a model of the proposed product — something fairly intangible. During the preference test, however, customers are exposed to a product they can touch, taste, or try. Much more information is presented to them, and because they are better informed their answers and reactions are likely to be better predictors of eventual market acceptance.

A preference test also provides clues to minor design improvements that can make the product even better. If the suggested design improvements turn out to be major, it's back to the drawing boards for a total redesign and more customer tests.

The final purpose of preference tests is to determine how and why the customer responds to the product. For this purpose, tape recordings of customer responses are invaluable. The words and phrases the customers use in their comments provide valuable hints about how the product should be communicated to the customer. The attributes or features that strike the customer first can be used in designing ads, brochures, or sales presentations.

Once the preference testing is complete, what does one do with the data? Can market acceptance or market share be estimated from the data? The following guidelines will help you to maximize the value of preference testing.

- Be careful not to "oversell" the product to the customer. If you make too forceful or biased a presentation, what you're probably measuring is how good a salesman you are, not whether the customer really likes the product. In the case of a new telecommunications product, for example, the project manager (who also was the product champion) conducted user tests and follow-up interviews himself. He was delighted with the consistently positive customer reaction. A second wave of tests and interviews, done by a third party, revealed much more negative results. It was found that the enthusiastic product champion had so oversold his product's benefits that he virtually coerced the respondents into positive responses. In another case, a focus group moderator managed to destroy the evaluation of a new office product. She became so positive about the product that she called for a show of hands in favor of the product. Her own hand was the first to go up. She made two mistakes: first, an open vote puts undue pressure on people: individuals are likely to be influenced by the more powerful members of the group. Second, the fact that the moderator voted — and voted first — produces an obvious bias in the group response. Fortunately, the session was videotaped, and management wisely disregarded the results of the evaluation.

- Be sure that the customer is sufficiently well-informed about the product to be able to judge it. This is a problem with innovative products. If the potential customer doesn't understand the product, its use, and its benefits, his or her responses won't mean very much. An "information session" held prior to the test should give the customer relevant facts concerning the characteristics, use, and purpose of the product if these are not immediately apparent.

- Be cautious in measuring price sensitivity in customer preference tests. A common ploy is to ask an "intent to purchase question" about a product priced at say, 99 cents. Then the question is repeated, and a price five cents higher is named. Not surprisingly, the proportion of "definitely would buy" responses goes down as the price goes up. This type of questioning is invalid. By quoting the first price as 99 cents, the interviewer has established a reference price that is likely to bias all subsequent price questions. Had half the respondents been presented one price, and the other half the higher price, the positive responses would have been much closer. The same problem arises when a list of possible prices is presented and the respondent is asked, "What's the most you'd pay for the product?" The reference range of prices influences the answer.

  If price sensitivity is an issue — that is, if you want to measure intent to purchase as a function of price — one price should be presented to one group of respondents, a second price to another group, and so on. Even with these controls, however, measuring price sensitivity is tenuous at best.

- Don't take "preference" and "intent-to-purchase" data literally. A 52 per cent preference level does not translate into 52 per cent of market share. The concerns relating to concept-test results also apply here. The results usually must be discounted to adjust for "yea saying," the lack of dollar commitment on the part of the buyer, and split purchases.

Some firms use the following rule of thumb: a minimum of 50 per cent of the target market must prefer the product, either "somewhat" or "very much," over the brand or make they currently buy or use. If the figure is below 50 per cent, the new product is in trouble. Another rule for a frequently purchased product is that market share equals the square of the preference level times a launch factor. Here the launch factor ranges from 0.2 (for a weak launch) to 0.6 (for a strong launch).

History is perhaps the best guide for translating preference and intent data into market-share estimates. This points to the need to conduct user tests for every new product and to build up a history of data.

- Interpret results of preference test of difficult-to-distinguish products carefully. For example, one cigarette manufacturer consistently obtained preference results on new products in the 45 to 55 per cent range — a respectable result, so management thought. Eventual market shares were disappointing, however. An investigation of the testing procedure was undertaken. It was found that when a preference test was conducted on cigarette A versus cigarette B, 40 per cent of the people preferred A, 40 per cent preferred B, and 20 per cent liked both equally. The catch was that A and B actually were the same cigarette! The point is that people will often indicate a preference where no difference exists, particularly in product categories where there are few product cues to help users distinguish between products. Thus, the preference results of 45 per cent obtained in the cigarette example aren't very meaningful: 40 per cent was by chance, and only 5 per cent was true preference. In product categories where cues do exist, however, preference results are more meaningful.[2]

Extended Trials

Extended user trials enable the customer to use a product over a longer time period, usually at his or her own premises. The customer's reactions and intents are thus likely to be based on better information. Extended tests are particularly appropriate for complex products, for products that require a learning period, and when it takes time for the customer to discover the product's strengths and weaknesses. An extended trial may also uncover product deficiencies not apparent in a short customer test or a lab test.

To undertake an extended trial, a sample of potential customers is identified and qualified (that is, they agree to participate). The product is then given or loaned to the customer. He proceeds to use it in his home, at work, or in the factory. A debriefing session is held with the user (either in a personal interview or by phone). The usual questions — interest, liking, preference, and intent — are posed. Probing questions can be asked about the product's strengths and weaknesses, its ease of use, its frequency of use, and suggestions for improvement. Often the results are unexpected. Witness the case of the wall telephone (described in chapter 3) that fell off the hook when a nearby door was slammed. There are other examples. A manufacturer of heavy equipment developed a prototype tree-harvesting machine. The unit was designed to fell trees with a knife-like action, strip the branches, cut the tree into sections, and load the sections onto on a carrying device. The unit was thoroughly tested by company engineers in nearby forests and pronounced satisfactory. The unit was then loaned for customer tests to a forest-products firm. All went well at the first site. When the unit was operating at a second site during rainy weather, serious product deficiencies became apparent. The combination of a certain soft soil (common in many forests) and wet weather caused the unit to become hopelessly mired in the mud. The test revealed that major changes were required in the traction design of the product — expensive changes, perhaps, but far less costly than the prospect of having dozens of the units stuck in the mud of forests around the world.

In the case of the milk-packaging system mentioned earlier, it was not until a test run was done with a potential customer that serious design problems with the product were uncovered. A prototype packaging machine ran perfectly in DuPont's development department, and a test at a dairy was then arranged. Everything went wrong. The packaging equipment was intricate and sophisticated; the company's development people also were sophisticated; but the dairy production people weren't. The machine was too sensitive and complicated for them to operate reliably. The coordination between the filling and tab-attachment operations was particularly unsuccessful. Once out of sync, tabs were suddenly being affixed to the bag an inch below or an inch above the prepunched hole. The result was thousands of leaking bags. Consequently, the equipment was simplified, troublesome operations were modified or eliminated, and in subsequent tests at the dairy the system operated smoothly.

User Tests: A Critical Ingredient

User tests and contacts, both during development and after the proto-type or sample is ready, often prove critical to the success of the product. Studies show that the customer test phase is significantly correlated with new-product success. Moreover, analyses of new-product failures reveal that in half the failures the customer test was poorly undertaken or skipped altogether.[3]

The objectives of customer tests usually include some or all of the following:

- to determine whether the product works well in actual use conditions (if not, what improvements are required?);
- to gauge whether the product is acceptable to the customer (why or why not?);
- to measure the customer's level of interest, liking, preference, and intent to purchase (and the reasons for these);
- to gauge price sensitivity — how preference and intent are affected by price;
- to determine those benefits, attributes, and features of the product to which the customer responds most strongly (information useful in the design of the communications strategy for the product).

Suggestion: The customer test phase is not a difficult step, nor is it unduly expensive. Given its pivotal role in identifying product deficiencies while there is still time to correct them, and in assessing likely market acceptance, it is recommended that customer tests be built into your game plan following prototype development and, if possible, during the product-development process itself. Remember: check with the customer, and check again — no surprises!

The Final Trials

By now, the product has been tested with the customer, and has been pronounced satisfactory. Minor design improvements have been incorporated. At the same time, the marketing plan for the product is coming together. (Chapter 8 is devoted to the development of a marketing plan; note that this marketing planning exercise gets underway in parallel with product development.)

Finally, the time is ripe to test the product, its production, and the launch plan under commercial conditions. For the first time you pull together all the elements of the marketing mix — product, price, advertising, promotion, sales force, etc. — and test their combined effect. At the same time, you produce a limited quantity of the product in a trial or pilot production run. The aim, of course, is to determine whether the strategy and programs as envisaged will generate the sales and profits you expect. If the answer is no, then you can choose between modifying the strategy and killing the project. It's still not too late to turn back.

Testing Market Acceptance

There are two possible ways to test the launch strategy. Both are essentially experimental. Both cost less and are less risky than a full-blown launch. Both serve to provide a fairly valid test of the launch strategy while leaving time for course corrections to be made before the launch. And both are reasonably good predictors of eventual sales or market share.

The first method is a pretest market — a simulated shopping experiment that has gained popularity among consumer-goods producers. The second is a test market or trial sell, which, although more expensive, has wider applicability for different types of products. Let's look at each in more detail.

The Pretest Market

A pretest market is a relatively inexpensive yet surprisingly useful method for predicting market share and sales from a new product. There are a number of commercial versions of pretest market studies offered by various consulting or market research firms. Examples include BASES, ASSESSOR, and TEMP.

Potential customers in a pretest market study are brought to a testing facility, where they are exposed to advertising for the new product or to a concept statement. In some approaches, the advertising is built right into a television show, and the consumer thinks he is there to view a pilot. Following the exposure, the consumer is given the opportunity to go on a simulated shopping trip through a dummy store. He is given coupons or credits and asked to select some merchandise. Of course, the new brand is displayed in the store, along with a variety of other typical store products. If the consumer chooses the brand under test, he is interviewed a few weeks later after using the product.

A pretest market study yields important information. First, the simulated shopping trip provides a measure of the effectiveness of the advertising and the package in generating sales. Second, information on product use, liking, and repurchase intent is obtained. The initial trial rate combined with the repurchase intent permits estimates of sales or market share. Finally, these techniques produce valuable segmentation data: demographics and other pertinent information about study participants are obtained, and a more exact definition of the target market is developed. Each of the commercially available pretest market techniques varies somewhat in terms of method, computation, and purpose. BASES is used primarily to predict Year 1 and ongoing volume, whereas ASSESSOR and TEMP predict ongoing market share.

Why have such techniques become so popular, particularly among consumer-goods producers? Cost is the big factor. A pretest market costs about $100,000; a test market can cost ten times that amount. Moreover, pretest markets are surprisingly predictive. Although the experiment is somewhat artificial — a simulated shopping trip, phoney money, a dummy store, etc. — experience has shown that the results are very close to the market share finally achieved after launch.

One major consumer goods firm estimates that pretest markets demonstrate an "accuracy factor" of plus or minus two percentage points. That is, if the predicted share was 10 per cent, then the actual share will be between 8 per cent and 12 per cent. Of 17 such pretest market results, in only one case were the results so far wrong as to lead to a bad decision in the project. The pretest market had predicted a substantially higher market share than was eventually realized; the product was launched, and failed.

Other reasons for using a pretest market include its timeliness (it doesn't take as long to set up and conduct as a test market); the depth of data provided (segmentation data on triers and repurchasers); and control. The last point merits mention. In a full-fledged test market, there are many variables beyond the control of those conducting the test. One of these is competitive activity. Stories are told of deliberate competitor interference: competitors cut their prices, increase promotional activity, and even sabotage displays, all in an effort to thwart the test market or invalidate its results. A pretest market, in contrast, is much more controlled: the store, the competitive brands on the shelf, and the displays are all within the control of the company conducting the test.

The one serious problem with a pretest market is its limited applicability. Pretests are typically limited to relatively inexpensive consumer goods — the kinds of products found on supermarket shelves. The dummy store, the simulated shopping trip, and the fake money are clearly inappropriate techniques to use with big-ticket consumer items or industrial products. For those products, a trial sell, or test market, is the best means of testing the proposed launch plan and product.

Test Markets

Test markets (or trial sells) are the ultimate form of testing a new product and its marketing plan prior to committing to the full launch. Of the testing techniques, a test market comes closest to testing the national launch strategy before it actually takes place.

A test market is essentially an experiment. As in any experiment, there are subjects, treatments, and a control group. A small representative sample of customers is chosen — they are the subjects. They are exposed to your new product and to the complete launch plan, which includes all the elements of the marketing mix. This is the treatment. (Several different treatments can be used on different groups to see which works best.) The control group is all people not exposed to the test market.

There are usually two reasons for conducting a test-market study. The most common objective is to determine the expected sales of the new product. A reliable forecast of future sales is critical to the final GO/KILL decision of the game plan. If the test market shows poor sales performance, the project can be killed or recycled to an earlier stage in the game plan, or steps can be taken to modify the launch strategy.

A second objective is to test two or more alternative launch plans using two different treatments to see which gives better results. This type of test marketing is less common. For one thing, it's clearly more expensive. Besides, the hope is that by the time you're ready to test-market, strategy questions will have been resolved. Nonetheless, in some cases the test market is used to decide which market strategy works best. The choice of an appropriate positioning strategy is one of those cases.

Some years ago, a food company planned to introduce a new instant-breakfast drink. The product had some taste advantages over competitors; it also was more convenient to prepare and store in the home. One possible strategy was to position the product as "a great-tasting breakfast drink"; the other was to position it as "a convenient, easy-to-prepare breakfast drink." Four test-market cities were chosen. Two were subjected to the "great taste" positioning strategy; the other two featured the "convenience" strategy. The "great taste" strategy consistently outperformed the "convenience" strategy, and the test-market results contributed to the decision to use the "great taste" strategy for the national launch.

Testing Industrial Products

Test markets can also be used for industrial products, in which case they're usually referred to as trial sells. A trial sell goes hand-in-hand with a pilot production run of the product. If a limited quantity of the product can be produced, samples can be made available to a handful of company salespeople in one or two sales territories for trial sell. The elements of the trial sell are as close to those of the actual launch as possible: the price, the advertising literature, the direct mail, and sales presentation are identical. The only difference is that national advertising and promotion cannot be used for a single sales territory. As with a consumer test market, negative sales performance in a trial sell will signal either a KILL decision or significant changes in the launch plan before the product is sold nationally.

Designing a Test Market

When the decision is GO for a test market, a number of decisions will have to be made to ensure accuracy and reliability of results.

Locations

The test market locations must be chosen. In the case of consumer goods, cities usually are selected; for industrial products, sales territories can be used. Locations should be chosen to be representative of the entire market. For consumer goods, this means representative in terms of demographics and other segmentation variables. Cities must be selected with the availability of appropriate local media in mind. For industrial goods, a representative sales territory means representative in terms of industry breakdown, size of buying firms, etc.

Two or more sites usually are selected for the test. If two alternative strategies are being tested, if the risks are high in the project, if uncontrollable variables are likely to be a factor, or if representativeness is a problem, then more than two sites probably will be required.

Execution

The test market itself amounts to an execution of the marketing plan, but only for the selected locations — a "mini-launch." All of the elements of the marketing mix should be as close to those of the final launch as is possible, including pricing, advertising, channels, and sales presentations. The duration of the test market must be established; tests can range from several months to several years, although shorter tests usually are preferred. Products with longer repurchase cycles necessarily mean a longer test-market period.

Measuring Results

Decisions must also be made on what data to gather. For consumer goods, warehouse shipments are a rough indicator of performance, but that figure also includes product already in the "pipeline." Sales to end users — retail sales via store audits — are the preferred measure. With industrial goods, sales to end users can be more directly measured, since the distribution channels tend to be shorter.

Some firms include end-user surveys in their test markets. Now that the product is actually in the hands of a customer, the time is ripe to obtain critical information. The task is to conduct follow-up interviews with users to find answers to some or all of the following questions:

- Who bought the product (demographics and other segmentation data)? Such information helps to confirm or refute the original definition of the target market.
- Why did he or she buy it? A knowledge of the "whys" leads to insights into the effectiveness of the communications and positioning strategies, and into buyer motivations and preferences.
- Did the customer like the product after he or she tried it? Why or why not? Such information is critical to a confirmation of the soundness of the product's design, features, attributes, and benefits.
- Would the customer repurchase the product? Answers to this question enable a determination of long-run market share to be made.

Incorporating an end-user survey into a test market provides far more information than the test market alone, which only measures sales results. The results of a survey can prove invaluable if the market-launch strategy needs modification or adjustment.

Identifying the User

Identification of the end users can be a problem for manufacturers of some types of goods. If follow-up interviews are to be conducted, provision must be made in the design of the test market to determine who should be interviewed. For industrial goods, the "who" information can be recorded as part of the sale, either through your own sales force or with the help of distributors. (The "who" information might include not just the purchasing agent, but also the individual or departmental user.) For big-ticket consumer goods, a mail-back in the guise of a warranty card provides this data. For smaller items, in-store intercepts can be used, or some form of redeemable mail-back can be included in the package.

To Test or Not to Test

Having examined the elements of a successful test market, we now move on to the most important decision: whether to undertake a test market at all. One emerging school of thought argues that test markets aren't worth the time, trouble, and cost. Test markets are expensive both in time and in money, and as the full launch draws nearer, lead-time over the competition becomes an increasingly valuable commodity.

Some pundits argue that Procter & Gamble had a sure-fire winner with the new double-dough process used in Duncan Hines Chocolate Chip Cookies. The baking process yielded cookies that were chewy on the inside but crispy on the outside, much like home-baked cookies. The product passed the concept tests, preference tests, and pretest markets with flying colors. But Procter & Gamble overdid the test market by carrying it on too long. By the time the product went national, competitors had developed similar products, and Procter's national launch wasn't as sweeping a success as originally predicted. It can be argued that with such positive results from earlier tests, and with lead-time so valuable, the test market in this instance could have been shortened in favor of a national roll-out.

Cost in money is another big factor. Test markets cost hundreds of thousands and sometimes even millions of dollars. The value of the information generated by a test market must be weighed against the cost of conducting the test market. In the cookie example cited above, a cost/benefit analysis probably would have ruled against a test market, particularly if the cost of lead time had been imputed.

Another argument against test markets is that they exemplify the "horse and the barn door" situation. By the time the test-market results are in, the door is being locked just after the horse has fled. Basically, the development budget has been spent; the product is fully developed, the creative work has been done, the packaging costs incurred, and the plant tooled up, at least for limited production. What's left? It's too late to make changes now — the time to have killed or modified the project was much earlier in the process. This argument is persuasive in cases where expenditures up to the point of commercialization (for example, development) are particularly high in relation to launch costs.

A final argument against undertaking a test market is the questionable validity of results. As noted above, much can go wrong with a test market. Many variables in the experiment are beyond the control of those conducting the test. Often those variables cannot be known until the test market is well underway, and by that time it's too late to do anything. In more scientific terms, the results of the experiment are found to be invalid because of noncontrollable variables. In one example, a firm tested several new products in different regions of the country. When asked how the test marketing was going, the product manager responded: "It's hard to tell — there's been a lot of competitive activity in those markets these last few weeks." That's a polite way of saying that competitors saw the new products being tested, and scrambled to mount a hard-hitting local drive against them — something they could easily do in a small market area, but not on a national scale. In the light of that competitive activity, how valid will the test-market results be? In another example, a major cosmetics firm tested a new shampoo in a major city. Unfortunately, that city was hit by a major labor strike in several key supermarket chains, which happened to be important retail outlets for shampoos. The test-market sales were obviously hurt by the strike, but how badly? No one knew, and the results of the test market were worthless.

Test markets or trial sells are not necessarily needless or wasteful. Give serious thought, however, to the pros and cons of undertaking such a test: test markets may not be an automatic or routine part of every new product's game plan.

A test market is useful when the uncertainties and the amounts at stake are high. A test market is warranted in the following types of circumstances:

- When there is still a high degree of uncertainty about the eventual sales of the new product as the launch phase approaches. When you've conducted all the appropriate tests but are still undecided about the product's market acceptance, a test market may be called for.
- When the horse is still not completely out of the barn — when there are many expenses yet to be incurred in the project before and during the full launch. If many expenditures remain to be committed in the project — for example, if a plant needs to be built or a production line retooled or set up; if an expensive national advertising campaign needs to be mounted; if a sales force needs to be hired and trained — then a test market can be used to provide valuable inputs to these final GO decisions. On the other hand, if the production facilities are in place, and the if launch is relatively inexpensive (that is, if future expenditures are low), the cost and time involved in a test market may not be justifiable.

Certain technical considerations must be borne in mind when deciding to go with a test market. Limited or trial production may not be practical for some products. As one manufacturer of telecommunications equipment put it: "For telephone handsets, there's no problem doing a test market. We can run a couple of thousand of these units down a quickly set-up production line quite easily. For major capital equipment, however, such as a new digital switch, the day we make our first production unit, that's the day we're in full scale production. There's no halfway."

Limited marketing in one or two cities, regions, or sales territories must also be possible. For goods that rely on electronic media, local print media, direct mail, local distribution channels, and personal selling, the marketing effort can be made to focus on one region. If national advertising and promotion vehicles are key to the product's launch, then a test market may be ruled out.

Suggestion: If the risks remain high as the project approaches launch, consider building into your game plan a final trial: a pretest market or a test market, accompanied by pilot production. A pretest market is recommended for consumer packaged goods as a cost-effective predictor of market acceptance. For other types of goods, however, a full test market or trial sell is really the only method of accurately predicting the final sales results.

Conclusion

Go into a test market with your eyes open. In too many cases test markets have been undertaken when they weren't really needed. Concept tests, preference tests, and pretest markets had been undertaken with positive results, yet the firms proceeded with the test markets anyway. When they were asked why, their answer was, "It's company policy." Given the predictive abilities of pretest markets and the problems and costs of test markets, clearly there are times when a test market can be safely omitted.

If you've done a thorough job on the up-front or pre-development activities and carried out usage tests, preference tests, and pretest markets, a test market may be unnecessary. The time to spot a bad product or a bad strategy is early in the game plan, not after the horse has bolted. In too many cases, unfortunately, a test market is a belated attempt to close the barn door. It's better to spend your time and money on up-front, "homework" activities.

If the decision is GO for a test market, remember that selection of locations, design of the test, and specification of what information will be gathered (and how it will be gathered) are critical issues, and require much thought. For example, think seriously about adding a buyer survey to the test market; for a small cost it provides the needed diagnostic insights that a simple test market doesn't yield. Too many test markets are badly designed, and others yield only limited information — yet they cost a fortune. Don't make that mistake in your test market.

The final evaluation decision — GO to full production and full market launch — is largely a financial one. Armed with the results of preference or end-user tests, test markets or trial sells, and pilot production runs, you can now make estimates of production and marketing costs, sales volumes, final prices, and profit margins with a high degree of confidence. Before the product moves to full-scale commercialization, a thorough financial analysis is essential. A discussion of the dozens of financial-analysis techniques available would take an entire book. Most of these techniques are fairly well known, and descriptions are available from a wide variety of other sources.[4] For now, keep these hints in mind when carrying out a financial analysis:

- Use a cash-flow technique, such as discounted cash flow (DCF) or capital budgeting. If you insist on relying on accounting or accrual methods for project evaluation — non-cash-flow techniques — you're headed for trouble. Those accounting techniques don't give true return rates, and cause many problems in implementation: should an item be expensed or capitalized, how long should the depreciation period be, and so on. In contrast, cash-flow methods yield the true return and avoid implementation problems; for example, it doesn't matter whether an item is an expense or capital item — the calculation is simply "cash in less cash out!"
- Recognize that money has a time value. Even at an inflation rate of zero, a dollar today does not equal a dollar next year. Use a financial method that discounts dollars that are earned or expended in future time periods. Again, the DCF method does this.
- Recognize that although your estimates of sales, costs, profit margins, and expenses appear valid today, they are based on guesses at future events. That means that you can never be certain about how unforeseen factors will affect your estimates.
- Build sensitivity analysis into your financial calculations. Take your "most likely estimate" first and perform your financial calculations. Then redo the calculations using pessimistic and optimis-

tic estimates for that same variable. In this way, you can find out how sensitive your return rate and your GO/KILL decision are to your original assumptions.

By now, the market and production tests have yielded positive results. Armed with those results, the final DCF and sensitivity analyses were carried out. The expected return clearly exceeds the minimum acceptable level, even with pessimistic estimates of key variables. So the decision is GO for commercialization. It's time for the final play of the game — into the market!

The Final Play: Into the Market

The Marketing Plan

The marketplace is the battleground on which the new product's fortunes will be decided. Thus, the plan that guides the product's entry to the market is a pivotal facet of the new-product strategy. In this chapter, we'll look at the factors involved in developing a marketing plan for your new product.

First, what is a marketing plan? It's simply a plan of action for new-product introduction or launch. It specifies three things:

- the marketing objectives;
- the marketing strategies; and
- the marketing programs.

The marketing plan itself is a document that outlines or summarizes your objectives, strategies, and programs. The marketing-planning process is a series of activities undertaken to arrive at the marketing plan. Much of this chapter will deal with the process of developing a marketing plan — setting objectives, developing marketing strategies, and formulating marketing programs.

Timing

When does the marketing-planning activity begin? This chapter occurs rather late in the book because the market launch is one of the final stages in the new-product game plan. Be warned, however: this is not to imply that marketing planning should be the final step prior to launch. If you leave it to the bitter end, you're likely to find you've done too little, too late.

During one of my investigations into how companies develop new products, I made an appointment to interview a senior executive in charge of new products. The company was a large manufacturer of heavy equipment. I arrived at headquarters for the interview, and was quickly directed to the engineering building several blocks away. "Mr. X, who's in charge of new products, is located in our Engineering Department," I was told. That should have been my first clue. During the interview, Mr. X spent several hours reviewing the development process. He focused almost entirely on the engineering, prototype-development, and product-testing phases. Finally I asked, "When do the other departments — manufacturing and marketing — get involved?" He replied, "Manufacturing? They enter the scene after the product is tested and we've developed a set of manufacturing drawings. And Marketing? Those sales fellows get involved as the product's getting ready for production — almost as the first unit comes down the production line." In subsequent conversations, it came as no surprise to learn that the firm's new-product performance was indeed dismal, and that many problems could be traced to a lack of an effective and carefully conceived launch plan.

Start Early

Marketing planning is an ongoing activity that occurs formally and informally throughout much of the new-product process. Informally, it begins during the first few stages of the game plan, right after the idea stage. By the time the project enters the concept stage formal marketing planning is underway. Exhibit 8.1 shows that the development of a marketing plan occurs simultaneously with product development to emphasize that a formal marketing plan should be in place long before the product is ready for market introduction or even for a trial sell.

Suggestion: Where in the new-product process does marketing planning occur in your firm? Does it begin, as it does in many firms, at the very end of the game plan? Or do you start marketing planning in parallel with the development of the product? If it's a matter of "too little, too late," why not incorporate the marketing-planning activities alongside the product-development phase of your game plan?

An Iterative Process

The marketing-planning process for a new product is an iterative one. The plan is not carved in marble at the early stages of the new-product process. Even the formal marketing plan that should be in place prior to product tests and trials is likely to be tentative. The first version of the plan probably will see many changes before it is

EXHIBIT 8.1.   Marketing Planning in the Game Plan

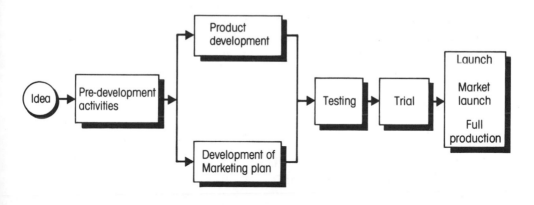

Adapted with permission from R.G. Cooper, "A Process Model for Industrial New Product Development," *IEEE Trans. Engineering Management* EM-30 (Feb. 1983): 2–11. Copyright ©1984 IEEE.

finally implemented in the launch stage. In short, there will be many times when you will rethink and recast your marketing objectives, strategies, and programs before implementation.

Setting Marketing Objectives

The marketing objectives that you specify for a new product must mean something. Why bother going through the aggravation of setting objectives at all? Objectives are part of a marketing plan for good reason.

The Role of Objectives

First, an objective is a decision criterion. When a manager is faced with two alternative course of action, she weighs the consequences of each action against her objectives. She then picks the alternative that comes closest to meeting those objectives. Thus, marketing objectives help managers make decisions about specific marketing actions.

Second, a common and well-understood set of objectives for a new product creates a sense of purpose — a goal for the team players to strive toward. The written objectives communicate this goal. This common understanding is critical, particularly if the new-product team is a large and diverse one. In too many new-product projects the players are on quite different wavelengths simply because the product's marketing objectives are not clearly specified, not written down, and not communicated. You've probably heard the saying, "Having lost sight of our objectives, we redoubled our efforts." The remark applies in too many new-product situations.

Finally, marketing objectives become a standard for measurement. Milestones or benchmarks are critical during the launch phase, when course corrections may be necessary. How will you know if you're on course if you haven't specified where you should be at any given time?

Good Objectives

What makes a "good objective"? Marketing objectives must

- set criteria for making decisions;
- be quantifiable and measurable;
- specify a time limit.

A typical objective might be expressed as: "To gain a leadership position in the market." This sounds laudable, but it's a poor objective. First, it is not useful as a decision criterion. Second, it isn't quantified. What does "leadership" mean? Does it mean "50 per cent market share or better," or does it mean "the highest market share among competitors?" And what does "market" mean? The whole market? Or a specific and narrow segment of the market? Third, because the objective isn't quantified, it can't be measured. For example, a year from now, after the product is out on the market, how will the product manager know if the product is meeting its objective? Finally, no time limit has been specified. Does the objective mean "in year 1" or "in year 10"?

There is a much better way to express the same marketing objective: "To obtain a 20 per cent unit market share in the owner-operator segment of the class 8 diesel truck market during year 2 in the market." Phrased in this way, objective is a guide to action. Alternative plans can be assessed on their likelihood of achieving a 20 per cent share; the objective is measurable and quantified; a time limit is specified; and market share can be measured during year 2 to determine whether the product is on course.

Typical marketing objectives for a new product should include some or all of the following:

- unit or dollar sales of the product by year;
- market share by year (be sure to specify the whole market or a segment and whether the share is measured in terms of units or dollars);
- product profitability — percentage margins, annual profits by year (dollars or percentage), and payback period.

Suggestion: Review several of your firm's past marketing plans for new-product launches. Take a hard look at the "marketing objectives" section of the plan. Did they establish good criteria for making decisions? Were they quantifiable and measurable? Did they specify a time limit? If not, strive for sharper objectives in future marketing plans using the list of typical objectives above as a guide.

Refining the Objectives

The process of setting objectives will involve iteration, or recycling. The setting of objectives is shown as stage 1 in the marketing-planning process in Exhibit 8.2. In practice, however, you must revisit this objective-setting stage a number of times as you move toward your final marketing plan.

At the early stages of the project, some rough numbers may be available that permit ballpark estimates of objectives. These early estimates may be little more than educated guesses, but at least you will have made your first attempt at setting some objectives for the product. As more and better information about the market, the product's expected advantages, and projected costs becomes available, the objectives will become better defined and more valid. Market studies, financial analyses, cost analyses, and other activities that are part of the new-product game plan are inputs to the constant refinement of marketing objectives. By the time the product is ready for launch, the marketing objectives will have undergone extensive changes from the first rough estimates made at the beginning of the project.

Realistically, marketing objectives for a new product represent a merging of what is desired and ideal and what is possible. In the final marketing plan, the objectives for the product — sales, market share, margins — and the forecasts for the product become one and the same.

The Situation Size-up

The situation size-up is a key facet of the marketing-planning process. Typically, it is shown as the step that precedes the development of strategies. In practice, however, size-ups are done often and at virtually every phase of marketing planning. A size-up is a situation analysis — pulling together the relevant information and asking, "So what? What does all this information mean to the development of my plan of action? What are the action implications?" Many situation size-ups are long, boring, and over-descriptive, and fail to

Exhibit 8.2.   Developing the Marketing Plan

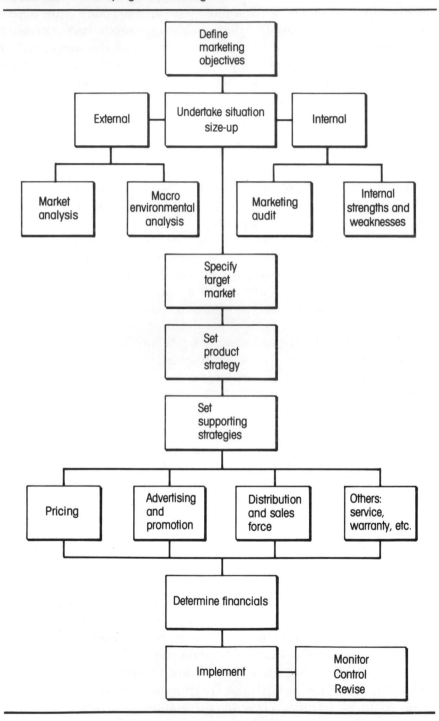

answer the question, "So what?" They begin with the heading "Background," then move to "Description of the Market," and so on. They're full of information and long on description, but short on action implications. Make sure that your situation size-up includes the pertinent information, but always tell the reader, "Here's what this means in terms of an action plan for our new product." The major areas that should be covered in a situation size-up in a new product marketing plan are shown Exhibit 8.2.

The Market Analysis

The market analysis lowers the microscope on the market for the new product. A good market analysis covers the important topics:

- *Market overview.* What are the quantitative and qualitative aspects of market size, growth, and trends?
- *Market segments.* What market segments are aimed at? How is each segment unique? What are the quantitative and qualitative aspects of their size, growth, and trends?
- *Buyer behavior* (in the segments in question). The who, what, when, where, why, and how of the purchase process are set out. Who buys? Who are the purchase influencers? What do the buyers buy, and when, and where? Why do they buy what they buy? What are their choice criteria and what are their preferences, wants, and needs?
- *Competition.* Who are the competitors? In which segments? What are their strengths and weaknesses? How good are their products? How does the customer rate their products? What are the competitors' strategies in pricing, advertising, and distribution? How well are they doing in market share and profitability? Why?

There are two points to remember: First, much of this market information will not be readily known at the outset of the new-product project. By the time the project is entering the product-development stage, however, market studies should have been undertaken and a thorough market analysis, with action implications, should have been completed.

The second point is that a good market analysis goes a long way toward charting a winning market strategy. If the market analysis lacks insight and information, the marketing plan probably will be vague and not very hard-hitting. A sound market analysis is the foundation upon which a winning launch plan is built. Don't skimp at this stage.

The Macroenviromental Analysis

A macroenvironmental analysis looks beyond the immediate marketplace for the new product. Trends and factors that lie outside the firm and the product's market that may impact on the market and product are analyzed:

- the economic situation;
- the political, legislative, and legal situation;
- demographic trends;
- social trends; and
- technological developments.

For example, when assessing at the economic situation in the case of a new home gardening product — say, a roto-tiller — one would look at, among other things,

- the GNP (as an overall indicator of wealth) and families' disposable income (current and projected);
- costs and prices of garden produce and inflation rates; and
- fuel-cost projections.

Under the "demographics" heading, one would look at the age breakdown of the population, population locations, and so on.

Several general questions should be asked for each trend category in the macroenvironmental analysis:

- What is the situation or trend?
- What is the timing of the situation or trend and how certain is it to occur? Is it here now or is it a "maybe and far in the future"?
- What are the implications of the situation or trend? Is it a threat or an opportunity? For example, what impact does the aging of the population have on the purchase of labor-saving home gardening products? For the design, positioning, and pricing of such products?
- What action is called for in light of the situation or trend?

The macroenvironmental analysis tends to be less concrete and focused than the market analysis, and some of the conclusions or action implications will be fuzzy and contradictory. Nonetheless, the analysis is a useful one to build into your marketing-planning effort. It doesn't take much time and effort, and on occasion some critical factors with a major bearing on the project are identified.

Internal Assessment

An internal assessment focuses on the company's internal strengths and weaknesses, particularly as they pertain to the project in question. A marketing audit typically is part of this assessment:

- Look at your sales force. Is it good, bad, or indifferent? What are its strengths and weaknesses? Will it be able to do a good job with the new product? If not, what should be done?
- What shape is your advertising in? Are significant changes and improvements required for the new product?
- Assess the status of the service department, the distribution or channel system, pricing policies, etc. What needs to be done to bring them "on line" for the new-product project?

The idea behind the marketing audit is to pinpoint marketing strengths and resources that one can build on and use to advantage in the new product. Remember: The shrewd strategist always attacks from a position of strength. An essential step in the strategy-development process, therefore, is to understand what your strengths really are, and to identify and correct weaknesses in the firm's marketing resources that could have a negative impact on the new product.

The internal assessment must also consider other facets of the company that will have a bearing on the launch plan for the new product. For example, you should be aware of the strengths and weaknesses of the manufacturing department — quality-control problems, availability of raw materials, people shortages, etc. Similarly, the strengths and weaknesses of other groups in the company, such as engineering, R & D, and finance, are equally critical. The object is to avoid being handicapped in your market launch by problems in other company departments.

Suggestion: Using past new-product marketing plans in your company as test cases, assess the "goodness" of the situation size-ups that were undertaken. Was the market analysis a good one? Did it touch on the points outlined above? Was the environment reviewed and assessed? Was a "strengths and weaknesses" audit undertaken? Most important, did the situation size-up point to action implications? If your situation size-ups have typically been weak, why not begin with an outline or map of what you want to see in such an analysis? Remember that a solid situation size-up makes the job of strategy formulation much easier.

Defining the Target Market

The importance of target-market definition is a key element of the protocol statement (see chapter 6). Clearly, one must have a precise definition of the target market before designing the product and developing the launch plan. In short, one must know "the object of one's affection."

How is a target market selected or defined? The first step is segmenting the market; the second step is selecting the appropriate segment to become the target market.

Segmenting the Market

Market segmentation is a popular topic among marketing strategists, and too complex to be fully discussed in this short chapter. Let's look at the highlights.

In the old days, economists spoke about "markets" as though they were relatively monolithic and homogeneous: "The market for X will behave this way or that way." Markets aren't homogeneous entities, however. They are people or groups of people buying things. No two people or groups of people are exactly alike, especially when it comes to their purchasing patterns. As consumers, we're all individuals: we have unique motivations, tastes, preferences, and desires. To treat all these different people or buying units as though they were painted with the same brush is naïve. Moreover, to try to appeal to those different customers with the same strategy — one product, one price, one communications approach — is counterproductive.

Market segmentation is the delineation of groups or clusters of people within a market such that there is relative homogeneity within each group and heterogeneity between groups. That is, the people within one cluster or segment exhibit more or less the same buying characteristics, but are quite different from the people in other clusters or segments. The company that develops a strategy tailored to a specific buyer or type of buyer is likely to be more successful than the firm that has only a single strategy in the market-place. Henry Ford's remark, "You can have any color as long as it's black," may have worked for the Model T and the early days of the automobile industry, but it fell flat once General Motors implemented a strategy of market segmentation in the 1920s: "A car for every purse and person." The idea behind segmentation is shown pictorially in Exhibit 8.3.

Exhibit 8.3. Different Ways of Viewing a Market

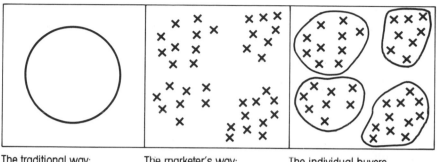

The traditional way:
The market is a
relatively homogeneous
entity

The marketer's way:
The market is composed
of individuals or
buying units, each
different from the other

The individual buyers
can be grouped together
into clusters or segments

This quick look at segmentation theory reveals that segmenting a market is a lot more difficult than picking a few convenient variables — age, sex, income (or, for industrial products, company size or Standard Industrial Classification (SIC) code) — and splitting the market into groups or categories. That is one method of segmentation, but the results usually aren't very helpful.

Bases for Segmenting Markets

Four broad categories of variables are useful in segmenting a market:

1. State of Being

Included in this category are familiar sociological variables such as age, sex, income, occupation, and the stage of the family cycle. The analogous variables for industrial goods are company size, industry classification (SIC code), and type of buying unit. Geography is another convenient variable: urban, suburban, exurban, and rural; regions of the country; and even regions of the world. The "state of being" variables are the easiest to use: they're familiar and easily measurable, and published statistics relating to the variables are usually available. Frequently, however, they don't yield useful segments. The statement, "We're introducing a new brand of after-dinner liqueur aimed at women" is nonsense. The target market "women" assumes that all women's tastes are the same when it comes to after-dinner liqueurs, and that's simply not true. Remember, seek clusters of people that are relatively homogeneous — when it comes to buying your type of product, the buyers within a segment should behave much like one another.

2. State of Mind

The variables in this category describe potential customers' attitudes, values, and lifestyle. This type of segmentation is called "psychographics." Consider the over-the-counter drug market. Psychographically, some people are hypochondriacs, some are skeptics, some are authority seekers. Each type is a different segment, to which a different product or different marketing strategy can be targeted.

3. Product Usage

Product usage segmentation looks at how the product is bought or used. The three major bases for usage segmentation are as follows:

- volume segmentation: the popular 80:20 rule applies for many markets — 20 per cent of the customers buy 80 per cent of the product. Buyers can be divided into categories of heavy user, light user, and non-user.
- loyalty: some customers are loyal to your firm, some are loyal to a competitor, and some move back and forth. The three different segments may deserve three different strategies.
- market factor: different people respond to different elements of the marketing mix. In many markets, there are price-sensitive buyers, quality-conscious buyers, convenience buyers, service-seeking buyers, and so on. These types of buyers are different segments and should be treated accordingly.

4. Benefit Segmentation

Perhaps the most useful method of segmentation for use with a new product is benefit segmentation. Benefit segmentation recognizes that people have different reasons and motivations for buying a product, and therefore they seek different packages of benefits. When purchasing a new car, some people are looking for basic transportation — a reliable, practical, low-cost, safe car. Others seek a prestigious high-performance car loaded with creature comforts. The heavy-duty truck study cited in chapter 6 was essentially aimed at identifying benefit segments. Several types of buyers, each seeking different "packages" of benefits in a new truck, were isolated.

Benefit segmentation is useful for new-product strategy. Using this approach, the target market defines the benefits that must be built into the new product. Usually, these benefits can be translated into specific product features, which aids the product-design process. The positioning and communications strategies also are largely defined by the benefit segment selected. Consider, for example, two target-market definitions for a new alcoholic beverage:

- Target market A: "The product will be aimed at white, middle-income, American women in the 20 to 30 age group."
- Target market B: "The product will be aimed at women seeking a mild, nonfattening, sweet-tasting, smooth beverage to be used in a relatively upscale social setting."

The definition of target market A is an example of demographic segmentation. The definition is too vague to be useful in designing the product or creating an ad campaign. The definition of target market B is an example of benefit segmentation. The product can almost be designed and advertised simply by reading the statement. Benefit segmentation has its drawbacks, however. It invariably requires extensive market research. That research is difficult — many intangible variables must be measured. It's much easier to measure people's ages or occupations than it is to measure the benefits they seek.

Suggestion: No doubt there is a great deal of discussion in your firm about market segmentation and about selecting the right target market when developing a marketing plan for a new product. But have you ever tried to segment the market based on the different packages of benefits customers want from a given product? Try benefit segmentation. You'll find it's a very powerful tool in designing a new-product strategy. Some market research will be required to determine what benefits are sought, and which people seek what benefits.

Selecting the Right Target Market

A segmentation analysis should yield a number of potential market segments. At the same time, different versions of the product may be conceived to suit two or more segments. As Exhibit 8.2 illustrates, when one thinks of market segments one also thinks immediately of how to target the segments — for example, what product benefits and features can be built into the product to suit it to a particular segment or segments. The next task is to select the appropriate target market from among these options. In the Gemini 2000 high-speed printer case cited in chapter 1, the computer-printer market could have been segmented into a number of submarkets. Some buyers had high-volume line printer applications; others wanted an occasional-use batch back-up line printer; some needed CRT hard-copy printers; others had plotting applications, either engineering or commercial. Each target market had a different set of needs for a high-speed printer, and probably required a modified version of the same basic printer. The task that Gemini management faced — and never successfully undertook — was to delineate the possible target markets, to define conceptually the appropriate printer for each, and then to select the best segment or segments as the main target market(s). Because management failed in the task, the final product and the launch were essentially shotgun efforts aimed at no specific application. Their "all things for all people" approach resulted in a product that satisfied no one.

Criteria for Market Selection

What are the criteria for selecting the target market (and product concept) from among a list of options? What criteria should Gemini management have used? Several straightforward standards can be established.

- *Segment attractiveness*. Which segment is the most attractive in terms of its market size, growth, and future potential?
- *Competitive situation*. In which segment is the competition the least, the weakest, or the most vulnerable?
- *Fit*. Where is the best fit between the needs, wants, and preferences of each segment and the benefits, features, and technological possibilities of your product?
- *Ease of access*. Which of the segments is the easiest for your company to reach in its selling effort, distribution channels, etc.?
- *Relative advantage*. In which segment do you have the greatest advantage over competitors in terms of product features and benefits, as well as other facets of your entry strategy? Note that "fit" and "ease of access" are not enough; they suggest mere adequacy. You must also look for areas in which you have a strong likelihood of outdoing your competitors.
- *Profitability*. It all boils down to profits! In which segment are you most likely to meet your sales and profits objectives?

Product Strategy

The definition of the product strategy — exactly what the product will be — goes hand in hand with the selection of the target market. Remember our discussion of the protocol: the protocol defines the target market and the product strategy. Target-market definition and product strategy, together, are the leading edge of strategy development, and are front and center in the development of the marketing plan (see Exhibit 8.2).

What is meant by "product strategy"? For a new product there can be three or four components to the term.

The Product's Position

Product positioning is a combination of market segmentation and product differentiation. "Position" in the marketplace means "how the product will be perceived by potential customers." The position is usually defined in terms of the key underlying dimensions by which customers perceive and differentiate among competitive makes. A possible positioning statement for the new truck described in chapter 6 would be: "A tough, long-lasting, durable, reliable, premium-priced vehicle designed for fast and easy maintenance and maximum up-time." In the Gemini case, the CRT hard-copy segment was a particularly attractive one. Had this segment been chosen, the positioning statement for the printer might have been: "A premium-priced, fast, cost-effective (low paper cost in this segment), quiet printer that produces clean, black-on-white, permanent, superior copies." The main competitive product in the CRT market at the time was a thermographic printer, which, although less expensive initially, used high-cost paper and produced poor copies using brown characters on a gray paper, which subsequently faded when exposed to sunlight.

Step 1 in defining the product strategy is the specification of the product's position — usually a sentence or two defining how the product will be positioned in the market and in customer's minds relative to competitive products and in terms of benefits offered. If you can't write down a clear, concise, and meaningful positioning statement, chances are you're headed for trouble. A fuzzy positioning statement is usually an indication of fuzzy thinking — no product strategy or, at best, only a vague notion of strategy.

Product Benefits

The benefits that the product will deliver to the customer should be delineated. Remember: a benefit is not a feature, although the two can be closely connected. A feature is part of the product's design — a physical thing. A benefit is in the eye of the beholder — some characteristic that is of value to the customer. For example, in the design of a new garden tractor, a benefit might be that the tractor can be operated easily by elderly people, including older women. Features that translate into this benefit might be a clutchless transmission, a hydraulic lift mechanism for the mower deck, and power-assisted steering.

In listing the benefits to be delivered, you'll probably find yourself repeating (and expanding upon) some of the phrases and words used in your positioning statement. It may seem a bit repetitive, but it's a worthwhile disciplinary exercise.

Features and Attributes

Step 3 is to translate the desired benefits into features, attributes, and product requirements. This step is likely to result in a much longer list of items, and one that gets very close to defining the product specifications. For example, in the truck case, if one benefit of the proposed vehicle was that it was "quick and easy to repair and maintain," then the corresponding list of features or product requirements might be:

- a quick-disconnect radiator; 2-4 butterfly bolts, several hose connections, and the rad is out;
- facility to drop the engine between the frame rails — the engine is out in half an hour;
- color-coded hoses and wiring with quick connecters; snap in, snap out for easy replacement.
- modularized electrics in the dashboard so that faulty modules can be pulled and replaced in minutes;
- 90-degree-tilt engine bonnet for easy engine access.

This three-step procedure — defining the position, listing the benefits, and itemizing the product features, attributes, and requirements — is a logical lead-in to the development of detailed product specifications, the fourth and final facet of the product strategy — an exact definition of what the product will be, and something tangible the development group can work toward. In some projects, detailed

Exhibit 8.4.   The Process of Developing a Marketing Plan

STAGES IN THE GAME PLAN

CORRESPONDING MARKETING-PLANNING
ACTIVITIES

STAGES IN THE GAME PLAN	CORRESPONDING MARKETING-PLANNING ACTIVITIES
1 IDEA Idea-generation Initial screening	
2 PRELIMINARY ASSESSMENT Market and technical assessments	FIRST CUT AT: - Marketing objectives - Size-Up of market - Defining the target market - Defining the product strategy

3 CONCEPT
Concept identification
Concept-generation
Concept test
Concept evaluation
(Protocol definition)
	DEFINE:
	- Target market
	- Product strategy
	  (Concept, positioning, benefits, features
	  and attributes)

4 DEVELOPMENT Product development Development of marketing plan	DEFINE SUPPORTING ELEMENTS: - Pricing - Advertising and promotion - Sales force, distribution, and service
5 TESTING In-house With customer	REVISE AND MODIFY PRODUCT (if needed)
6 TRIAL Trial production Test market	FINALIZE PRODUCT DESIGN REVISE AND MODIFY SUPPORTING ELEMENTS; FINALIZE PLAN
7 LAUNCH Full production Market launch	IMPLEMENT MARKETING PLAN MEASURE/CONTROL/MODIFY PLAN

product specs may not be possible at this point, and creative solutions by the development team may be required.

Marketing Planning and the Game Plan

You will have noticed that the marketing-planning process, outlined in Exhibit 8.2, closely parallels the new-product game plan. Indeed, if Exhibit 8.2 is superimposed on the game plan shown in Exhibit 8.4, key marketing-planning steps correspond to the various stages of the game plan. For example, the first few steps of developing the market-ing plan — setting objectives, undertaking a situation size-up, defin-ing the target market, and determining the product strategy — cor-respond to the three up-front stages of the game plan. Target-market definition and product strategy are critical facets of the concept stage, and constitute the protocol, the key step just preceding product development. Thus, the marketing-planning process gets underway much earlier in the game than many people might imagine.

The Supporting Elements

By now the leading edge of the marketing plan has been developed — the target market and product strategy. The top of the pyramid in Exhibit 8.2 is in place. Now come the support strategies, the remain-ing blocks in the structure. These are the elements of the marketing mix that will support the product launch. Let's have a quick look at the more critical ones.

Pricing

There are two ways to price a new product: cost-oriented pricing and market pricing. The following are examples of each approach. A company develops a new variable-speed electric motor. A detailed cost analysis is done, and the variable costs of producing the product are calculated at $500 per unit, largely made up of labor and mate-rials. Factory overhead brings the cost up to $700, and corporate over-heads add another $100. The full cost is $800 per unit. Corporate policy is to achieve a 20 per cent profit margin on the selling price, which will be $1,000 ($800 divided by 0.8). This is a simple example of cost-oriented pricing; the illustration of pure market pricing is even simpler. When asked to explain how his prices were set, the owner-manager of one heavy-equipment manufacturing company selected a large unit on the loading dock: "Take that 300 HP unit. Here's how I price it." He went to his catalogue shelf, took out his main competitor's price book, found the equivalent model, and declared: "Their price is $100,000. Mine is $95,000. Simple as that!" When he was asked what his manufacturing cost was, he replied, "Does it matter? Either I sell it at $95,000 or I don't sell it at all. As for costs, the unit costs somewhere between $60,000 and $75,000 to manufacture. The exact cost is irrelevant — it still wouldn't change my price."

Was he correct? Were manufacturing costs truly irrelevant to his pricing decision? The illustrations given here represent extremes, and both pricing methods have their drawbacks. In the first example, the pricing policy did not consider the realities of the marketplace. It's fine to say that at a price of $1,000 you'll make your target of a 20 per cent margin. But who says you'll sell *any* units at $1,000? The best strategy might be to take a lower margin per unit and sell more units. Cost-oriented pricing fails to consider key questions, such as "What is my product worth to the customer?" and "What price will the market bear?"

The example of pure market pricing, although less elegant in methodology, was likely closer to the truth. The marketplace does determine the product's worth. In this case, "worth" or "value" was judged relative to the price of a competitor's product, and price is simply a reflection of the product's value to the customer. By ignoring costs, however, the manager may not have chosen the price that will yield the best profit. For example, suppose he could sell twice as many units by selling at $85,000 instead of $95,000. If the variable cost of producing the unit is $60,000, he will be money ahead to drop his price to $85,000 (assuming that production capacity was available, and that competitors didn't react). If his variable cost is $75,000, however, dropping the price from $95,000 to $85,000 would be a breakeven proposition, and probably not worth the trouble.

How does one go about pricing a new product? It is difficult to generalize, but there are some basic guidelines.

1. What Is the Product's Target Market and Positioning Strategy?

Before you reach your pricing decision, both the target market and the product's positioning strategy must be specified. For example, if the product is aimed at a "niche" market, one with specialized needs, and if the positioning is a highly differentiated one, in essence you have a mini-monopoly situation: for that target market, you become the one and only product. A premium-price strategy is the route to follow. Conversely, if the product is not well differentiated from competitive products, and if the target market is served by others, a competitive pricing policy is appropriate.

Just in case you're tempted to enter the market on a "low-ball" price basis (that is, using a low price as a means of gaining market share), remember that price is the easiest strategy for a competitor to counter, while a product advantage may take years to catch up with. Similarly, an advantage gained through a clever promotional program, a unique distribution effort, or a massive personal selling campaign may force the competitor to play catch-up ball for months or even years. In contrast, a price advantage is usually temporary: it can be countered tomorrow morning with a simple telex to all salespeople, dealers, and distributors announcing a similar price cut.

It does make sense to use price as a leading weapon when you have a sustainable and real cost advantage: when your costs are truly lower than competitors' by virtue of product design, low-cost access to raw material, cheaper labor, or higher production volumes. Unfortunately, most firms are not in the position of being low-cost producers, especially in the case of a product new to the company. A low-price policy also makes sense is when it is part of a long-term strategy; the decision is to sacrifice profit to "buy" market share for the future, or to open a window for future new products. (These topics are discussed in more detail below.)

2. What Are the Other Strategic Issues?

There are a number of other strategic issues that may impact on your pricing decisions.

Skimming versus penetration. One school of thought argues that a pricing policy that yields low selling prices, high volumes, and low production costs is desirable. The profit per unit is low, but bigger profits come from volume. Usually, a larger investment in production facilities is required. The idea is to dominate the market through penetration pricing, and reap the long-term rewards of a leading market share. An assessment of your own strengths and weaknesses, your financial capacity and risk averseness, the slope of the learning or experience curve (costs versus production volume), the price sensitivity of the marketplace, and possible competitive reactions will dictate whether such a policy is a viable option.

The high-volume, low-price policy has many adherents. The PIMS studies (Profit Impact of Market Strategy) point to market dominance and high market shares as the key to profitability.[1] Similarly, the BCG model (the Boston Consulting Group's approach to strategic planning) relies heavily on experience curves and on gaining market share as the key to having a portfolio of "star products."[2]

The alternative to high volumes and low prices is a skimming policy. The new product is aimed at the market segment for which the product has the most value, and which will pay a premium for it. Profit per unit is high, but volumes are lower. Investment in production facilities is also lower, so the risk is often lower. While the product may never dominate the entire market, it may dominate the one segment and prove very profitable.

A combination of the two strategies is also possible. A skimming strategy is implemented to start with, attacking the high-value market segments. The initial risk is low. Should the product gain acceptance, and when the investment is partly paid back, then a penetration policy is adopted: increase production, drop prices, and go for dominance across the entire market. Timing is critical. The shift must take place before competitors invest in the development and production of similar products.

Corporate strategy. The new product's pricing must be established in the light of the corporate strategy. The new product is not a "stand-alone" item; it is part of a grander plan. For example, senior management may have decided that a specific market or product category is top-priority, and will commit significant resources, at a loss if necessary, to gain a foothold in the market. The new product may be the advance landing party that will sustain heavy losses while paving the way for more profitable future entries. Normal pricing practices may give way to larger issues. For example, several years ago Daimler-Benz made a decision to enter the North American heavy-truck market. Its two pronged strategy was based on price. First, it purchased an independent local truck maker, Freightliner, and pro-ceeded to wage a price war in the United States and Canada. (Some industry experts speculate that the firm is prepared to commit — to lose — up to $1 billion dollars to gain a foothold in the market.) Second, it moved a low-cost, lighter-duty truck manfactured in South America into the United States, again on a price basis. Relative to Daimler-Benz's European trucks, the design is obsolete, but the price is attractive. These low-priced products are the landing party; high profits will be temporarily sacrificed in exchange for effective market penetration.

3. What Is the Product's Value?

All new-product pricing boils down to an assessment of the product's value or worth to the customer. Value, like beauty, is in the eye of the beholder — the customer, in this case. Value is subjective; percep-tions vary with the buyer. The price is objective, set by the seller. Ideally, the price accurately reflects the product's value.[3]

Because two people can look at the same product, however, and judge it to have a different value, the first question to ask is: value to whom? If you've done an effective job in defining the target market, that question will have been answered. The next question is: what is the product's value or worth? In assessing value to the customer, one usually looks at what the customer's options are. If similar products are available to the customer, then your product's value is simply the price of the alternative to the customer, plus or minus a bit, depending on the advantages of your product, service, delivery, reputation, etc., relative to the competitors'. In pricing in highly competitive markets characterized by relatively homogeneous products, start with competitive prices and work upward or downward from there.

If your product is significantly different from what is now on the market, it is often possible to impute a value by comparing the product's worth relative to the product the customer is now using to solve his or her problem. For example, some years ago, a firm introduced a new building material aimed at builders of prefab homes. The product was a 4-by-8-foot panel of very thin bricks attached to a backer sheet and was designed to replace conventional brickwork on the exterior of a prefab home. The product's main advantage was that it could be factory-installed, thus eliminating on-site labor. The product was an innovation, so there were no directly competitive products upon which to base a pricing policy. The customer's alternative was conventional bricklaying at the job site; the value of the new product to the customer was calculated based on those material and labor costs.

When the new product has economic benefits to the customer — for example, measurable cost savings, as in the brick-panel example above — a value-in-use can be calculated and used as a standard for the product's value to the customer.

4. If in Doubt, Research the Customer.

Often, the only way to assess accurately the product's value to the customer is through market research. This research can be combined with the concept test or the product tests. There are several ways to gauge product worth and price sensitivity:

Ask the potential buyer, "What is the maximum price at which you would buy this product?" Naturally, you'll get different answers from different people, but plotting "percentage of respondents (cumulative)" versus "maximum price" gives an indication of price sensitivity (or price elasticity).

In measuring the intent to purchase, expose different groups of people to different prices. For example, divide your research sample into three groups. Present the product concept to group A at price 1; to group B at price 2; and to group C at price 3. Measure the intent to purchase, and plot "the percentage who said definitely yes" versus "price level." Again, this curve gives an indication of price sensitivity.

Use trade-off analysis in your concept tests. Different versions of the same product are presented to the respondent. The product can be varied along a number of possible dimensions, of which price is one. Sophisticated data analysis is used to determine the utility (or worth) of different features or attributes to the user.

A test market or trial sell also can be used as an experiment to test different price strategies.

5. What Is the Contribution Profit per Unit?

Contribution profit is the selling price less variable costs per unit (direct labor, materials, sales commissions, etc.). The manager who used his competitor's price book in the example above was right in at least one respect. The place to start a pricing analysis is at the top line, not the bottom line, of a profit analysis. The first question to ask is: at what price will the product sell? Based on the assessment of the product's value to the customer, there will probably be a range of possible prices. For example, assume that a manufacturer of consumer packaged goods wants to set a price for his new product. The competing product is sold at a retail price of 99 cents. The manufacturer's new product offers unique benefits to one significant segment of the retail market, so he believes that it can be sold at anywhere from 99 cents to $1.39. The retail margin is 20 per cent of the retail price, so manufacturer's revenue per unit of sale is the retail price less 20 per cent. For example, at a retail price of 99 cents, the manufacturer's revenue per unit is 79.2 cents. Always begin your price-setting calculations with a range of *selling prices*; if you base the selling price on your costs, you may find yourself locked into a "cost-plus" mentality — a deadly error.

Next, subtract the variable costs per unit. Variable costs are those that vary directly with output or production volume. They include items such as material, labor, the variable part of overhead costs, and some marketing expenses such as sales commissions. Don't include fixed costs — those that are incurred regardless of the production or sales volume of the product. Fixed costs typically include items such as depreciation on production equipment, management costs, light, heat, and rent. The variable costs of the product in the example above — largely ingredients, but also including packaging, processing, and variable distribution costs — are $0.708 per unit. When these variable costs are subtracted, what remains is the contribution profit per unit, shown in Exhibit 8.5. At a retail price of 99 cents, there is a contribution of about 8.4 cents per unit; for every unit sold at 99 cents retail, the manufacturer is "money ahead" by 8.4 cents. That 8.4 cents is contribution — it goes to cover the fixed costs. What's left over is profit.

Exhibit 8.5.   Pricing Using Contribution Profit Analysis

Retail price per unit	$0.99	$1.09	$1.19	$1.29	$1.39
Wholesale price per unit (revenue)	0.792	0.872	0.952	1.032	1.112
Variable costs per unit	0.708	0.708	0.708	0.708	0.708
Contribution profit per unit	0.084	0.164	0.244	0.314	0.404

From the contribution-analysis table in Exhibit 8.5, we can see that the manufacturer will have to sell three units at 99 cents to make the same profit he would by selling one unit at $1.19. In other words, he will only need to obtain a 10 per cent market share if he sells the product at $1.19, while he must have a 30 per cent share to make the same profit at 99 cents.

The contribution-profit analysis gives significant clues as to the direction in which our pricing policy should move. Compare the top line of Exhibit 8.5 with the bottom line. A 10 per cent price hike doubles the contribution profit! It's hard to believe, but true. The contribution profit per unit is sensitive to changes in price; just how sensitive it is can only be determined by such a contribution-profit analysis. Remember to use variable costs only. They're the relevant costs.

6. What Is the Total Contribution at Various Prices?

Now it's time to start thinking about possible sales volumes at the different prices. Often the sales volumes need only be educated guesses. If adequate research has been done to gauge the product's worth to the target customer and to obtain an idea of price sensitivity, then this calculation becomes more valid.

Based on end-user and concept tests, the manufacturer in our example estimates a reasonable drop-off in sales units at higher prices in the example. If he can sell 100 million units at 99 cents retail, volumes are expected to be 85 million units at $1.09, 60 million units at $1.19, and so on, as shown in the second line of Exhibit 8.6. The third line shows his contribution per unit, taken from Exhibit 8.5.

The total contribution profit is found by multiplying relative volumes at the different prices by the contribution per unit (line 4 of Exhibit 8.6).

What's the best price according to this analysis? The contribution

Exhibit 8.6. Contribution Profits at Various Prices

Retail price per unit	$0.99	$1.09	$1.19	$1.29	$1.39
Expected sales volume (millions of units)	100.	85.	60.	40.	20.
Contribution profit per unit	$0.084	$0.164	$0.244	$0.314	$0.404
Contribution profit (thousands of dollars)	$8,400	$13,940	$14,640	$12,560	$8,080

table (in Exhibit 8.5) pointed to a higher price than 99 cents retail. Further, the product's benefits, intended target market, and positioning strategy also suggested a premium-pricing strategy. The results of the analysis in Exhibit 8.6 show how much of a premium can be attached.

By reading across the bottom line of Exhibit 8.6, we see that the total contribution rises at first: the manufacturer makes a higher and higher contribution per unit, but sells fewer and fewer units. The total contribution peaks out at a price of $1.19. When the price goes higher than $1.19, the total contribution starts to drop. At a price of $1.39, for example, the contribution per unit is the highest, but fewer units are sold on which to make that contribution profit. Based on this analysis, a price of $1.19 might be the logical retail price. If market share or sales volumes were also important objectives, then the lower price of $1.09 might be chosen: it yields marginally lower total contributions, but sales are higher by about 40 per cent.

Where do fixed costs fit into this pricing exercise? In a nutshell, they don't! Fixed costs are relevant in the decision to enter a business. Once you are in, however, variable costs are relevant for pricing decisions. Still skeptical? Repeat the calculations in Exhibit 8.6 and subtract fixed costs of, say, $1.5 million, from the bottom line. Does the inclusion of fixed costs change the pricing decision?

7. Promotional Pricing

There can be a big difference between the ongoing or "normal" price and the introductory price of a new product. The pricing calculations, market research, and positioning strategy may all point to a premium price. But management may lose heart and feel that the price is too high to induce initial sales. A lower, less than optimal, price may be chosen. In the example above, one might have been tempted to elect a parity price of 99 cents, just to "remain competitive." If obtaining introductory sales is a major problem, don't sacrifice a well-conceived pricing strategy to do so. An introductory "promotional" price can be used to induce those first sales.

Promotional pricing can take many forms. For consumer goods, it can be coupons, a cents-off deal, or a company rebate. For industrial goods, a simple explanation that an introductory price is being offered to the first customers to buy the product will suffice.

There are several advantages to using an initial promotional pricing strategy. First, the "normal price" is retained for the long term. The customer is aware that the usual price is $1.19, but since the product is new it can be had for a limited time for 99 cents — a real deal. Second, the positioning strategy remains intact. If the product really is a differentiated and superior one, then it *should* be priced higher. The customer sees the "normal" price of $1.19, which reflects the product's worth or positioning. If a normal price of 99 cents had been elected, then the customer would see the product as just one of many — a regular-quality, regularly priced product. From the customer's perspective, price is an indicator of quality!

Finally, it's difficult to justify a price hike to the customer after the product is priced at 99 cents. With an introductory offer, however, making the transition from the introductory price of 99 cents to the normal price of $1.19 is easy, and is actually anticipated by the customer.

Suggestion: Although pricing is one of the most critical decisions of the new product's marketing strategy, all too often the pricing decision is handled in a sloppy fashion. Moreover, too often managers get locked into a "cost plus" mentality — prices are based on costs rather than on what the product is worth to the customer. In this section, seven key points to remember have been highlighted. Use this list the next time you face a new-product pricing decision. The result will be a much more thoughtful approach to pricing, and usually a better decision.

Advertising: Getting the Message Across

A company can have the best product in the world and sell it at a fair price. If no one knows about it, however, the battle is lost. The product's virtues must be communicated to its target market. Advertising is one effective communication tool.

Normally, the advertising plan is developed by an advertising agency or an in-house advertising or graphics department. The new-product manager is often tempted to wash her hands of the advertising function — to subcontract this facet of the marketing plan, and assume that "those advertising folks will handle it."

This attitude is wrong. An effective advertising campaign begins with the new-product manager. While the details of the media plan and the development of the "creative" (the artwork and copy) may be the task of others, the communications strategy itself is the product manager's responsibility. There are some simple "before" and "after" steps that can be taken to ensure a more effective advertising effort for the new product.

The "Before" Steps

Before meeting with the advertising agency (the term "agency" is used to denote either an outside or in-house group), here's what to do:

1. Specifying the Advertising Objectives

Advertising can do many things. It creates awareness, knowledge, and understanding. It can shape attitudes and create a desire or a preference for a product. In the case of direct marketing, it can even create a sale. Advertising can do all these wonderful things — for a cost! Before talking to the agency, pin down *what you want your advertising to do for you.* The product's advertising objectives should be specific and quantifiable. Some examples:

- To create an awareness, within three months of launch, among 50 per cent of the defined target market that Product X is now available.
- In six months, to have 30 per cent of municipal water engineers, buyers, and consultants aware that a new water pipe is corrosion-resistant and has double the life of a traditional ductile iron pipe.

The role of advertising in the total selling effort must be decided before specifying detailed advertising objectives: how much of the communications job will advertising do, and how much will be done by the sales force?

2. Specifying the Target Market and Positioning Strategy

A good advertising person will insist on knowing these in detail. Without a clear definition of the target market, how can he design a media plan? And without a positioning strategy, how will he know what the message is to be?

3. Describing the Target Market

The new-product manager must provide as much detail as possible on the target market and how it behaves: demographics, locations, occupations, etc. Other types of segmentation may have been used, such as benefit or volume segmentation. That's fine for most of the elements of the marketing strategy, such as product design, pricing, and so on. But remember, the advertising industry, and certainly the media-plan facet, still relies heavily on traditional segmentation variables in the choice of appropriate media. For consumer products, readership and viewership are still reported in terms of age, sex, income, etc., and for industrial goods by industry and by the reader's occupation or position.

4. Communicating the Product to the Agency

The agency should study the product thoroughly before embarking on campaign development. You can help by providing as much detail as possible on how the product works, how it is used, and what its benefits, features, and attributes are.

David Ogilvy, one of the gurus of the advertising industry, preaches to members of his industry about the importance of a painstaking study of the product prior to creative development. He cites the example of his Rolls-Royce advertising: how was the true "meaning" of a Rolls-Royce automobile to be translated into a print ad? After three weeks of reading about the car, Ogilvy hit upon the statement, "At 60 miles an hour the loudest noise comes from the electric clock." That became the headline message of this now-famous campaign.[4]

With these four key steps in place, it's time to turn the advertising development over to the agency. The agency will devise a media plan — which media will be used, the frequency and timing of appearance, and the budget allocation — and the advertisement itself. When the agency presents the results of its efforts, the new-product manager must once again become a key player in development of the advertising plan.

The "After" Steps

The review and approval of the proposed advertising plan is next. The steps are as follows:

1. Reviewing the Media Plan

The essential question is whether the proposed media plan will reach the target audience with the desired frequency. The plan should specify the expected reach and frequency of the campaign: how many potential customers the campaign will reach, who these people are, and how often they will receive a message.

First, look at each medium recommended by the agency, and in particular at the readership or viewership of that medium. Then compare that with the defined target audience and the advertising objectives. Second, determine how often the target customer will be hit with a message. The choice of frequency is largely based on experience. A good rule of thumb is that it takes three impressions for a person to get a message. A mere awareness of your product is likely to require a minimum frequency of three; more ambitious objectives — knowledge of product benefits or features, liking, or preference — will require a higher frequency.

2. Reviewing the Creative

Does the message back the product's position? Does it get across the product's benefits to the reader or viewer? An ad may be extremely creative and artistic, and may even win awards. The real purpose of an ad, however, is effective communication of the product. Don't feel shy about asking probing questions and critiquing the ad's potential effectiveness as a communication piece.

3. Running Tests

If the advertising budget for the product is large enough, you may want to test the ad. For example, the pretest market procedure described in chapter 7 performed a test of an ad's effectiveness. Another method is to measure customers' preferences for products on a list before and after viewing the ad. The advertising agency will be able to design appropriate testing procedures.

For low-budget campaigns, the testing will be done on a smaller scale and at a lower cost. Advertising for industrial products, for example, may take the form of a brochure, a direct mailing, or trade journal advertising. You can obtain feedback on your proposed ad by exposing it to a handful of customers, either individually or as part of a focus group, to measure its suitability and effectiveness.

4. Assessing the Worth of the Objectives

Now comes the tough question. The proposed plan will include a budget. Review the advertising budget with the original objectives in mind. You can then decide whether to accept the costs as reasonable in light of the objectives, or to back off on some of the objectives — perhaps they were too ambitious to start with.

5. Building in Measurement

The only way to know whether the advertising plan is achieving its objectives is to build in some techniques of measurement. Decide, with the agency, how advertising effectiveness will be measured. This will usually involve a market research study. For example, if one of the advertising objectives was that "in six months, 30 per cent of municipal engineers will know that our pipe has double the life of the competitor," then plan to take a representative sample of municipal engineers in six months. Ask them what they know about the new kind of pipe. If significantly less than 30 per cent of the sample don't know that it has double the life of the competitor, then the ad didn't achieve its objective.

Commercial services are an alternative to market research: such services regularly measure viewership or readership of ads in various media.

Suggestion: You've probably heard someone remark that 50 per cent of advertising dollars are wasted. The problem is that no one knows which 50 per cent! It's true that advertising is very much an art. Well-informed advertising decisions can be made however. Use the "before" and "after" steps and rules outlined in this section; be tough on the advertising people, and see if you can't improve this important element of the launch effort.

Sales Force Decisions

For the majority of new products, sales force decisions will be straightforward: the product will be sold by the company's existing sales force and/or through its existing distribution system. In the project-selection process, several important questions will have been asked:

- Will the product be sold to a market we now serve?
- Will the product be sold to our existing customers?
- Can the product be sold by our sales force and/or via our present distribution system?

If the answers to those questions were "yes" — and they are for most new-product projects that pass this screen — then the sales force plan boils down to tactical issues:

- training the sales force in the selling of the new product;
- providing the sales force with the appropriate selling aids;

- devoting effort to the new product (for example, developing a call plan to introduce the product).

For some new products, however, the use of the existing company sales force and/or distribution system may be inappropriate. If changes or additions to your sales force are to be made, two important questions must be answered:

- What is the nature of the selling job for the new product?
- Is the nature of the selling job compatible with the talents, training, and the way the current sales force (or distributor) operates?

An example: A manufacturer of scientific lasers and instruments marketed a product line of nitrogen lasers and other light sources in a low price range, typically $1,000 to $5,000 per unit. Its "sales force" consisted of a network of manufacturers' reps throughout North America, Europe, and Japan who called on scientific accounts. As scientific products went, the sale was a relatively simple one. The product was easily understood by buyers; it was easy to explain and demonstrate on-site by salesmen; it was a low-risk purchase item; and the client's purchase decision was typically quick and uncomplicated.

A new product introduced by the firm represented a significant departure. Unlike the simple lasers, this new optical instrument was a system — a sophisticated unit, priced at about $100,000. It could not be demonstrated on-site; many people were involved in a lengthy purchasing process; considerable explanation of the system's features was required; and it was a high-risk purchase decision for the customer. Naïvely, the company moved the new product through its usual sales force system. The product manager handled local sales and was available to the reps for back-up. Not surprisingly, the reps failed to perform. The selling task was so different from that connected with the usual products the reps handled that they were simply unable to cope with the new product.

In making sales force decisions for your new product — whether to use the existing sales force, hire a new sales force, or use a third party (a middleman) — the decision rests on a few critical factors:

- the fit between the nature of the selling job for the new product and the talents, training, and operating methods of the sales force — how they sell now;
- the degree of control over the selling effort that you need to exercise;
- the relative costs of each option, and whether those costs are fixed or variable.

The Other Elements

The main elements of the launch are now in place: the product and target market definition, the price; advertising; and the sales force. The remaining elements, not discussed in detail here, are physical distribution, promotion, and service and warranties. Those must also be built into the launch plan. They are critical to new-product success, but they tend not to be unique to new products. Moreover, they are typically already in place in most firms.

The Final Step: The "Financials"

The financial statements are an integral part of any launch plan. They cover two topics:

- what the plan will cost to implement (the budget); and
- what the plan will achieve (sales and profits projections).

The "financials" are detailed pro-forma profit-and-loss statements for the new product for year 1, year 2, year 3, etc. — in essence, a financial plan for the project.

Most new-product managers are not overly fond of bean counters — the accountants and financial experts. They are thought to be better at telling us where we've been than they are at predicting where we're going — at dealing with the past than with the future. The new-products game is very much future-oriented, but this is one time when a solid financial analysis and a financial plan are essential.

The financial plan is important for several reasons. First, it serves as a budget for the new product — an itemized accounting of how much will be spent, and where. Second, the financial plan is the critical input for the final GO/KILL decisions as the project moves closer and closer to full launch and commercial production. The expected return from the product can be computed from the financial plan. Finally, the financial plan provides benchmarks. These benchmarks are critical to the control phase of the launch plan — making sure that the new product is on course. A launch plan should also include contingency plans for actions to be taken if the results deviate from the expected course.

In developing a launch plan, and particularly for the first attempt or first iteration for a specific new product, the financial plan is often the acid test. Any major discrepancies in strategic thinking are discovered and dealt with at this point; for example, there may be major differences between the objectives and the financials. The original sales and profit objectives set out at the beginning of the planning exercise may be miles apart from the sales and profits spelled out in the financials. Or, there may be inconsistencies between costs of achievement and expected results. The financials often reveal that the costs of implementing the plan are simply not warranted by the results the plan will achieve.

The existence of such discrepancies is no surprise. The marketing-planning exercise is very much an iterative one. This was the first attempt — a roughed-out, tentative plan. Now, go back to the beginning of the planning exercise, and start again — the refining process. Rethink the objectives; redo the size-up; reformulate the action plans; recalculate the financials. These iterations or recycles take time and effort. That's why it is a good idea to start the marketing-planning effort so early in the new-product process.

As the product moves closer and closer to the launch, and with each successive iteration and refinement to the plan, the launch plan starts to crystallize. And if the homework, tests, and trials have been properly executed, it should be a matter of clear sailing into a successful launch, with another winner on your hands!

The Long Term: What Markets, Products, and Technologies?

Win the Battle, Lose the War?

What if...

- What if a manager implemented the game plan — the new-product process from idea to launch as outlined in the last six chapters?
- And what if she religiously followed the game plan?
- And what if the steps were executed well?
- And even assuming luck was on her side...

would the result be a steady stream of successful new products with a high-profit impact on the firm? Not necessarily. One ingredient is missing, and that ingredient makes the difference between winning individual battles and winning the entire war.

The key ingredient is the product innovation charter (PIC).[1] The PIC charts the entire strategy for a firm's new-product program. It is the master plan, and it is the essential link between the product-development program and the firm's corporate strategy.[2]

How does the PIC fit into the game plan? Some strategic models show it as preceding the idea-generation stage, as though the PIC were simply one stage of many in the game plan. But the PIC is more than that; it overarches the game plan and influences every stage of the new-product process. The role of the PIC is shown pictorially in Exhibit 9.1.

What Is a Product Innovation Charter?

A PIC is a strategic master plan that guides the new product game. But how does one define or describe a new-product strategy? The term "strategy" is widely used in business circles today. The word is derived from the ancient Greek word meaning "the art of the general." Until comparatively recently, its use was confined to the military. In a business context, strategy has been defined as "the schemes

Exhibit 9.1. The Product Innovation Charter and the New-Product Game Plan

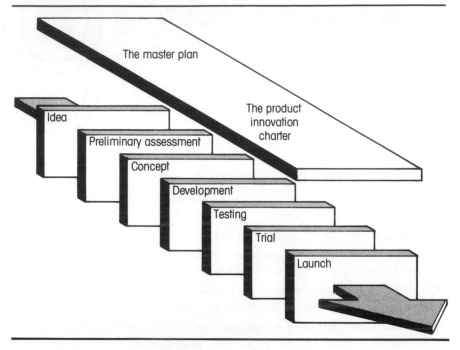

whereby a firm's resources and advantages are managed (deployed) in order to surprise and surpass competitors or to exploit opportunities."[3] More specifically, strategic change is defined as "a realignment of the firm's product/market environment."[4] Strategy is closely tied to product and market specification; Corey identifies market selection and product delineation as the two key dimensions of corporate strategy.[5]

Product-innovation strategy, while closely related to corporate strategy, tends to be more specific. Typically, it contains two key elements.

- The PIC specifies the objectives of the game, the role that product innovation will play in helping the firm achieve those objectives, and how new products and product innovation fit into the company's overall plan. Statements such as "By 1990, 30 per cent of our corporate sales will come from new products that we will develop and launch in the next five years" are typically found in the PIC.
- The PIC specifies the arenas in which the game will be played. That is, it defines the types of markets, market applications, technologies, and products that the new-product program will focus on.

Why Have a PIC?

Developing a PIC is hard work. It involves many people, and especially top management. Why, then, go to all the effort? Most of us can probably name countless firms that do not appear to have a master plan for their new-product program. How do they get by?

Doing Business Without a PIC

Running an innovation program without a PIC is like running a war without a military strategy. There's no rudder, there's no direction, and the results are often highly unsatisfactory. On occasion, such unplanned programs do succeed, largely owing to good luck.

A new-product program without a PIC will inevitably lead to a number of ad hoc decisions made independently of one another. New-product and R & D projects will be initiated solely on their own merits and with little regard to their fit into the grander scheme. The result is that the firm finds itself in unrelated or unwanted markets, products, and technologies.

Objectives: The Necessary Link to Corporate Strategy

What types of direction does a PIC give a firm's new-product program? First, the objectives of a PIC tie the product-development effort tightly to the firm's corporate strategy. New-product development, so often taken as a given, becomes a central part of the corporate strategy, a key plank in the company's overall strategic platform.

The question of spending commitments on new products is dealt with by defining the role and objectives of the new-product program. Too often the R & D or new-product budget is easy prey in hard economic times. Research and development tends to be viewed as "soft money" — a luxury. But with product innovation as a central facet of the firm's corporate strategy, and with the role and objectives of product innovation firmly established, cutting this budget becomes less arbitrary. There is a continuity of resource commitment to new products.

The Arenas: Guiding the Game Plan

The second facet of the PIC, the definition of arenas, is critical to guiding and focusing the new-product efforts (see Exhibit 9.2). The first step in the game plan is idea-generation. In chapter 4, a proactive

Exhibit 9.2.   The Choice of Arenas Guides the New-Product Process

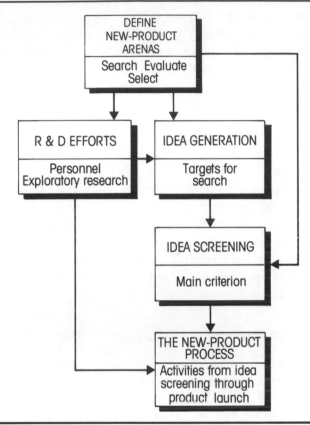

Reprinted with permission from R.G. Cooper, "Strategic Planning for Successful Technological Innovation," *Business Quarterly* 43 (Spring 1978): 46–54.

idea-generating system was proposed as both practical and desirable. But where does one search for new product ideas? Unless the arenas are defined, the result is a scattergun search effort, undirected, unfocused, and ineffective.

The second step of the game plan is idea screening. The first criterion for this early GO/KILL decision is whether the proposed new product falls within the company's mandate for its new-product program. This usually translates into "Is this the kind of market, product, and technology that we as a company have decided is fair game for us?" Without a definition of fair game — arenas — good luck in trying to make an effective screening decision!

The definition of arenas also guides resource and manpower planning. If certain markets are designated top-priority arenas, then the firm can acquire resources, skills, and knowledge to enable it to attack those markets. Similarly, if certain technologies are singled out as arenas, the company can hire and acquire resources to bolster its abilities in those fields. Resource building doesn't happen overnight. One can't buy a sales force on a moment's notice, and one can't acquire a critical mass of key researchers or engineers in a certain technology at the local supermarket. Putting the right people, resources, and skills in place takes both lead time and direction.

Where's the Evidence?

So far the argument for a PIC, although logical, appears somewhat theoretical. One can't help but think of all those companies that have made it without a grand strategy. So where's the evidence in support of a PIC? There isn't much, unfortunately. The reason is that few studies have investigated the role and impact of an innovation strategy. Most of the business research into product innovation has focused on the individual product as the unit of analysis — for example, on what makes a new product a success — rather than on a company's entire new-product program.

The studies that have looked at firms' new-product strategies have a clear and consistent message: new-product strategies at the firm level are critical to success, and some strategies clearly work better than others.

• Booz Allen & Hamilton's study of new-product practices in corporations found that "successful companies are more committed to growth through new products developed internally" and that "they are more likely to have a strategic plan that includes a certain portion of growth from new products."[6] The authors of this study go on to explain why having a new-product strategy is tied to success:

> A new product strategy links the new product process to company objectives, and provides focus for idea/concept generation and for establishing appropriate screening criteria. The outcome of this strategy analysis is a set of strategic roles, used not to generate specific new product ideas, but to help identify markets for which new products will be developed. These market opportunities provide the set of product and market requirements from which new product ideas are generated. In addition, strategic roles provide guidelines for new product performance measurement criteria. Performance thresholds tied to strategic roles provide a more precise means of screening new product ideas.[7]

• The PIMS studies (Profit Impact of Market Strategy) considered new-product strategies, but in only a peripheral way.[8] The studies looked at why certain business units were more profitable than others, and attempted to link profitability to the market strategy elected. Research and development spending and product-quality level were two of many strategy variables considered in the studies, and both were found to be connected to profitability.

• Nystrom and Edvardsson studied a number of industrial-product firms, and identified how various new-product strategies were tied to performance.[9] Strategies emphasizing the synergistic use of technology, a responsive R & D organization, and an externally oriented R & D effort were generally more successful. While the study was limited to a handful of strategy dimensions, the message is clear that strategy and performance are closely linked.

• In a recent study, I looked at the performance impact of product innovation strategies in 120 firms.[10] This study is one of the few investigations undertaken to date that considers many strategy dimensions, and how strategy was tied to performance in a large number of firms. The overriding conclusion was that product innovation strategy and performance are strongly linked. The types of markets, products, and technologies that firms elected and the orientation and direction of their product-innovation programs had a pronounced impact on the program's success and profitability. Strategy really does count.

How Product-Innovation Strategies Affect Performance

What are the secrets of a successful innovation strategy? To answer this question, I observed the product-innovation strategies of 120 firms, measured on 66 strategy variables. These strategy variables described the types of markets, products, and technologies that the firms elected for their new-product programs, and their direction, orientation, and commitment to this program. The performance of the firms' new-product programs was measured on 10 different scales. The conclusions, set out below, are based on concrete data and the results of a scientific investigation, not on wishful thinking, conjecture, or speculation. They will prove useful in the formulation of a PIC.

Conclusion 1. There is a strong connection between the new-product strategy a firm elects and the performance results it achieves. New-product strategy and performance are closely connected. The underlying hypothesis of the study — that strategy leads to performance — was strongly supported. Five different strategy types or scenarios were uncovered, and each was associated with a different performance level and type. One strategy, called "Strategy Type B" (a balanced strategy), yielded excellent results overall, and in a convincing manner.

New-product success is not solely a matter of good fortune, or even of being in the right industry. Admittedly, firms in certain types of industries — growth industries, technologically developing industries, and high-technology industries — on average achieved better new-product performance. This comes as no surprise. The point must be made, however, that the new-product strategy elected had a pronounced and independent effect on performance. That is, the types of arenas selected, and the firm's direction and commitment to the program, all helped to determine performance.

The implications of this strategy-performance link are critical to the management of a firm's new-product efforts. The existence of this link points to the need to define clearly the firm's new product strategy as a central and integral part of the corporate plan. The development of a PIC becomes a pivotal management task.

During discussions with managers, I found that the great majority either lacked a written PIC, or admitted to having only a superficial plan. Many firms did not even have quantifiable objectives for their new-product programs. For example, managers often did not know key performance results of their programs, and had to do considerable digging to answer questions on straightforward objectives-and-control gauges, such as percentage sales by new products or success, fail, and kill rates.

Suggestion: New-product strategy pays off. If your firm does not have an explicit written PIC, complete with measurable objectives and specification of arenas as a guide to your firm's new-product efforts, now is the time to begin developing one.

Conclusion 2. On average, new-product development programs performed well. Product development programs have a much better performance record, on average, than previously assumed. Over the years, some startling "statistics" surfaced, such as the often quoted "90 per cent of all new products fail" and others. Crawford's study helped to dispel many of these myths by showing that most of the reported figures were based on speculation, on personal claims, or on studies of questionable merit.[11]

Consider our findings on new-product performance. Overall, the average success rate for developed products was 67 per cent. Only 17 per cent of launched products failed in the marketplace, and another 16 per cent were killed prior to launch. Admittedly, these figures do not include the many projects that were killed partway through development, after substantial amounts of money and time were spent on them.

New-product performance was also positively rated on other performance criteria. New products introduced in the five years prior to the study represented, on average, 66.5 per cent of the current sales of the firms. On a number of scaled questions that gauged performance — contribution to corporate sales and profit objectives, overall profitability, success versus competitors, and overall success — companies scored on the positive side of the ratings.

The disconcerting evidence, however, was the wide variation in those performance measures:

- Success rates for developed products ranged from zero to 100 per cent. More than 25 per cent of the firms had a 50-50 success rate (or worse) on launches.
- Percentage sales by new products (the percentage of current corporate sales that came from products introduced in the last five years) also ranged from zero to 100 per cent. Almost 40 per cent of the firms reported that 20 per cent or less of their sales were from new products.
- On the scaled ratings of performance (zero-to-10 scales), the average rating was six to seven; answers showed high variations among firms, however, from lows of zero to highs of 10.

The point must be made again: Most new products do succeed, and most new-product programs contribute in a major way to the sales and profits of the firm. These results do much to dispel the negative myths that product development is a luxury only a few firms can afford, that R & D spending is a questionable expenditure yielding low returns, and that only a fortunate few succeed in the new-product game.

Suggestion: Does your firm suffer because too many people believe the myths — that the new-product game is too high-risk, that the payoffs are low, and that one is better off avoiding the game? If so, use the concrete evidence in the studies to help dispel the myths and to put new-product development in a more positive light in your firm.

Conclusion 3. All firms do not follow the same new-product strategy; five separate strategy types exist.

Type A: the *technologically driven strategy*, the most popular (used by 26.2 per cent of firms), featured a technologically driven sophisticated approach to product innovation. The program lacked a strong market orientation, and there was little fit, synergy, or focus in the types of products and markets exploited. The markets tended to be unattractive ones. (The details of each strategy type are shown in Exhibit 9.3.) This strategy generally led to mediocre performance

Exhibit 9.3.   The Five Strategy Scenarios

A TECHNOLOGI- CALLY DRIVEN	B BALANCED STRATEGY	C TECHNOLOGI- CALLY DEFICIENT	D LOW-BUDGET, CONSERVATIVE	E HIGH-BUDGET DIVERSE
Poor product fit and focus	Strong product fit and focus	Product differential advantage: quality and superiority	Production and technological synergy	Poor program focus
Targets low-potential, low-growth markets	Avoids competitive markets	Defensive orientation	Low R & D spending	High R & D spending relative to sales
Technologically sophisticated, oriented, and innovative	Seeks high-potential growth markets	Serves needs new to firm	Low-product differential advantage: impact and features	Targets competitive markets
Low marketing synergy	Avoids serving needs new to firm	Strong program focus	Serves needs new to firm	Targets markets new to firm
Weak market orientation	Strongly market-oriented	Low technological sophistication, orientation, and innovativeness	High product fit and focus	Premium-priced products
Avoids competitive markets	Technologically sophisticated, oriented, and innovative	Market-oriented	Low product differential advantage: quality and superiority	Poor production and technological synergy
Low product-differential advantage: quality and superiority	Strong program focus	Poor production and technological synergy	Avoids dominant competitor markets	Low technological sophistication, orientation, and innovativeness
Strong program focus	Premium-priced products	Avoids markets new to firm	High marketing synergy	Seeks high-potential, growth markets
Avoids serving needs new to firm	Avoids custom products	Targets dominant competitor markets	Low technological sophistication, orientation, and innovativeness	Weak market orientation
	Product differential advantage: quality and superiority	High R & D spending relative to sales	Targets highly competitive markets	Poor product fit and focus
	Product differential advantage: impact and features	Avoids premium-priced products	Avoids premium-priced products	Avoids serving needs new to firm
	Avoids markets new to firm	High marketing synergy		Product differential advantage: impact and features
High impact; low success rate; poor profitability	Top performers; best on every performance guage	Very poor results	Good success rate; low-impact program	Very poor results

Adapted with permission from R.G. Cooper, "The Performance Impact of Product Innovation Strategies," *European Journal of Marketing* 18 (1984): 1–54.

results — a program that failed to meet the firm's new-product objectives, a high proportion of cancellations and failures, and an unprofitable new-product program. The strategy did have a high impact on corporate sales, however. In sum, the technology-driven strategy is a technologically aggressive, moderately high-impact program, but costly, inefficient, and plagued by failures because of a lack of focus and a lack of marketing orientation and input.

Type B: the *balanced strategy* (15.6 per cent) featured a technologically sophisticated and aggressive program, a high degree of product fit and focus, and a strong market orientation. New products were aimed at attractive high-growth, high-potential markets where competition was weak. New products were premium-priced and featured a strong differential advantage: high-quality products that performed a unique task and met customer needs better, and products with a strong customer impact that offered unique features and benefits to the customer. Not surprisingly, this strategy led to the best results: the highest percentage of sales by new products (47 per cent versus 35 per cent for the other firms); the highest success rates of developed products; a higher profitability level; and greater impact on corporate sales and profits.

Type C: the *technologically deficient strategy* (15.6 per cent) lacked technological sophistication. Firms using this strategy pursued products that were a poor fit with the existing technology and production base of the company. They lacked an offensive stance, and attempted to serve market needs that they hadn't served before. Predictably, their results were dismal.

Type D: the *low-budget conservative strategy* (23.8 per cent) featured low R & D spending and new products with minimal differential advantages ("me too" products). The programs were focused and highly synergistic, tending toward a "stay close to home" approach. New products matched the firm's production and technological skills and resources; fit into the firm's existing product lines; and were aimed at familiar and existing markets. In spite of their lack of spending, firms using this strategy achieved moderately positive results: a high proportion of successes and low failure and kill rates. The program was profitable, but yielded a low proportion of sales by new products and had a low impact on corporate sales and profits. This conservative strategy resulted in an efficient, safe, and profitable new-product program, but one without a dramatic impact on the corporation.

Type E: the *high-budget diverse strategy* (18.9 per cent) was essentially a non-strategy. It featured heavy spending on R & D, but in a scattergun fashion; there was no direction, no synergy, no focus, no fit. The firms attacked new markets and new technologies, and used unfamiliar production technologies — a clear case of not sticking to their knitting. The firms were tied with the Type C firms as the worst performers.

Suggestion: Take a step back for a moment, and consider your firm's or division's new-product program. Is there a strategy at all, either implicit or explicit? If so, which of the five strategy scenarios comes closest to describing your firm's approach? How do your perform-ance results compare?

Conclusion 4. One strategy — Type B, a balanced strategy — yielded exceptional performance results. The strategy that outper-formed the others called for a balance between technological sophis-tication and aggressiveness and a strong market orientation. The performance results of the firms that elected the balanced strategy were dramatically better than those of the rest of the firms. The Type B firms were significantly higher than the other four strategy groups in terms of

- program success vis-à-vis competitors' programs;
- program importance in generating corporate sales and profits;
- meeting new-product program objectives; and
- the overall success of the program.

The Type B firms tended to excel in sales by new products (47 per cent versus 35 per cent for the other firms); success, failure, and kill rates (72 per cent success rates versus 66 per cent for the other firms); and program profitability.

Several characteristics distinguished the high performers from the rest. First, they had a strongly market-oriented and marketing-dominated program; they were technologically sophisticated, oriented, and aggressive; and they were highly focused. Second, they selected familiar markets that were noncompetitive, high-potential, and high-growth, with needs that the firm had served previously. Third, they developed new products that fit into their current prod-uct lines and that were closely related to each other. The products had two types of differential advantages: product quality and supe-riority, and unique features and benefits with a high customer impact. The products tended to be premium-priced, but not limited-scope, custom products.

The orientation of these firms' programs serves as a guide to other companies. Type B companies were the only firms to achieve both a *strong market orientation* and a high level of *technological sophistication and aggressiveness.* These firms possessed technological prowess comparable to that of many other firms, yet they based their new product program on the needs and wants of the marketplace. Their new-product ideas were derived from the marketplace; a proactive search effort was made for market-need identification; a dominant marketing group was involved in the new-product process; and the entire process had a strong market orientation.

The types of products and markets that these top performers selected (outlined above) were unique, and can serve as a guide to other firms in the selection of arenas as part of the firm's PIC.

Finally, the balanced strategy yielded positive results independent of the characteristics of the firm's industry or the firm itself. Industry growth rate, technology level, and technological maturity of the industry all affected performance, but the most important factor was the choice of the right strategy. Moreover, the balanced strategy gave consistently positive results regardless of firm or industry. This winning strategy is also a universally applicable strategy.

Suggestion: Compare your firm's explicit or implicit innovation strategy to the one elected by the Type B strategists. Do you share the same orientations? Do you select the same types of markets? Do you develop similar types of products? Go through the list of distinguishing characteristics of these firms, and see how you rate on each item. This exercise should shed light on your strategic strengths and weaknesses.

Conclusion 5. Adopting some, but not all, of the elements of the winning strategy is not sufficient. Certain elements of the balanced strategy can be found in other strategy types. None of those types performed nearly as well as Type B, however. For example, strategy D, the low-budget conservative approach, shared certain elements with B, namely a good product fit and focus. The Type D firms also possessed a high degree of technological and marketing synergy between their new-product projects and the firm's resource base. The result was second-best, however, and far short of the winning strategy. In particular, while the success, failure, and kill rates of new products were positive, the low-budget, technologically unaggressive strategy simply lacked the R & D commitment and technological prowess of strategy B. The result was a low-impact program — a case of winning the battle, but losing the war.

Strategy A firms adopted a technologically aggressive stance, like the winning Type B firms. But they lacked a market orientation, developed products that were a poor fit with their marketing resources, and tended to target low-growth, low-need markets. The result was a moderately high-impact program, but one with a poor success, failure, and kill rate.

The conclusion is that a technologically driven and dominated strategy, on its own, is wrong. Equally wrong for most firms is a conservative, stay-close-to-home approach to new products. The most successful strategy is one that marries technological prowess, a strong market orientation, and a high degree of fit and focus: Strategy B.

Conclusion 6. A low-budget conservative strategy yields fairly positive results, especially for some types of firms and industries. The low-budget conservative approach was one of the most popular strategies. It was found to work well only for some types of firms. Companies adopting the conservative strategy featured new-product programs that:

- had a high degree of technological and production synergy;
- had a low level of R & D spending; and
- created products with the fewest differential advantages — "me too" products in their customer impact and features.

To a lesser extent, the same companies' programs

- had a low level of technological sophistication and orientation;
- created products with little differential advantage in product quality and superiority;
- had a high degree of product fit and focus;
- featured products priced lower than competitors';
- were targeted at markets synergistic with the firm, but involving needs new to the firm; and
- were targeted at highly competitive markets, but markets with no dominant competitor.

On average, firms adopting the low-budget conservative strategy achieved positive results in program profitability (returns versus expenses) and new-product success rates. The end result was a low-impact program, however, with a lower-than-average percentage of sales from new products (31 per cent versus 38 per cent for all the other firms). For certain types of firms the low-budget conservative strategy worked particularly well. Companies with strengths in marketing (strong sales force, channel system, advertising, and market research skills) and firms in technologically mature, slower-growth industries performed extremely well by adopting the strategy. Sound performance was restricted to their profitability and success rates, however; the program was still low-impact.

One conclusion is that firms that possess certain distinctive competences — marketing power, for example — might rely on those strengths as the key to moving relatively ho-hum new products to the market. But the results for Type D firms were still inferior to the balanced-strategy firms that faced similar markets and had similar strengths. Moreover, for firms lacking key strengths or facing developing and higher-growth industries, the conservative strategy typically yielded results inferior to the balanced-strategy approach. Further, while a conservative strategy may work well for some firms and over the medium term, if the firm's markets or technologies change dramatically the firms are caught in a vulnerable position — victims of the "product life-cycle trap."[12]

Suggestion: Is your company a Type D firm — facing mature markets, but with key marketing strengths in your company? Have you elected the low-budget conservative approach to new products? If so, and if you're typical, the results of your new-product program are probably adequate. But they could be even better if you adopt the balanced-strategy approach.

The Big Message

This extensive study of firms' innovation strategies shows clearly that successful product innovation begins with the determination of a new-product strategy. The strategy-performance link uncovered in the study is solid evidence of the need for strategy definition. The study provides many insights into the ingredients of a successful innovation strategy — the types of objectives that are both reasonable and measurable, and the criteria that are useful in the selection of arenas (the kinds of products, markets, and technologies that successful firms elect). Finally, the practices of the top performers — a focused program with a strong market orientation married to technological sophistication — provide strong hints about how the new-product game should be played.

Developing a Product Innovation Charter: Setting Objectives

A few months ago, I boarded an early-morning flight on a major airline. The captain began his announcement: "Welcome aboard flight 123 en route to... ah..." There was a long pause. The pause was punctuated by laughter and wisecracks from the passengers; the captain didn't know where the flight was going! Fortunately, within 30 seconds, he remembered our destination. If he hadn't, that plane probably would have emptied. Who would stay on a plane where the captain didn't know his destination? Many of us, however, seem content to stay on board new-product programs that have no destination. Defining objectives for a product-development program is essential; most of us accept that premise. Yet our strategy study revealed that many firms actually lacked written and measurable objectives for their firms' innovation programs.

What types of objectives should be included in a PIC? First, the objectives should be measurable. Second, they should tie the new-product program to the total corporate strategy. Finally, they must give the new-product team a sense of purpose and help them make decisions. In deciding upon a reasonable set of new-product program objectives, you may want to consider some of the following types.

Role Objectives

One type of new-product objective focuses on the role that the new-product effort will play in achieving corporate objectives. Examples include:

- The percentage of company sales in year 5 that will be derived from new products introduced in that five-year period. (Five years is a commonly accepted time span in which to define a product as "new.") Alternatively, one can speak of absolute sales — dollars in year 5 from new products — rather than relative sales or percentages.
- The percentage of corporate profits (gross, contribution, or net) in year 5 that will be derived from new products introduced in that five-year span. Again, absolute dollars can be used instead of relative profits.
- Sales and profits objectives expressed as a percentage of corporate growth. For example: 70 per cent of growth in company sales over the next five years will come from new products introduced in this period.
- The number of new products to be introduced. There are problems with this type of objective, however: the products could be large-volume or small-volume ones, and the number of products does not translate conveniently into sales and profits.

Note that the objectives listed above can be broken down by business unit or company division, or even by new-product type; for example,

50 per cent from extensions and modifications, 30 per cent from new items in existing lines, 20 per cent from new lines and/or new-to-the-world products.

The specification of role objectives gives a strong indication of just how important new products are to the total corporate strategy. The question of resource allocation and spending on new-product efforts can then be more objectively decided.

Performance Objectives

A second type of objective deals with the expected performance of the new-product program. Such objectives are useful guides to managers within the new-product group. Examples include:

- success, failure, and kill rates of new products developed;
- number of new product ideas to be considered annually;
- number of projects entering development (or in development) annually; and
- minimum acceptable financial returns for new-product projects.

Many of the performance objectives flow logically from the role objectives. For example, if the firm wants 70 per cent of sales growth to come from new products, how does that figure translate into number of successful products, number of development projects, success, failure, and kill rates, and number of ideas to be considered annually?

Setting these objectives is no easy task. The first time through, the exercise is often a frustrating experience. Yet these objectives are fundamental to developing an innovation strategy. Without them, it's difficult to arrive at an innovation strategy, not to mention a logically determined R & D budget figure.

Suggestion: Step 1 in developing a PIC is defining objectives. Start with the role objectives listed above, then move to the performance objectives.

Defining the Arenas

Defining the arenas is the next phase of developing a PIC; what business areas should the firm's new-product efforts focus on? Remember that program focus was found to be a key ingredient in a successful new product program. Further, arena definition is important for idea-generation, for screening new-product projects, and for manpower and resource planning.

There are two steps to defining the target arenas. The first is developing a comprehensive list of possible arenas. The second is paring the list down to the most appropriate arenas.

What Is a "Business Arena"?

How does one go about describing a "business arena"? Defining a business has been the topic of much business writing. In his famous article, "Marketing Myopia," Theodore Levitt suggests that we define the business in terms of the markets and needs the firm now serves.[13] He urges managers to remove their blinders and to define the business not in terms of products, but in terms of markets and needs served. Levitt cites the example of U.S. railroads: had the railroads defined their business as "transportation of people, goods, and 'things'" instead of as "railroading," they might well be in a variety of more promising businesses today, including road, air and sea transport, hotels, telecommunications, etc. In Levitt's view, arenas are simply "market/need" areas, and the main criterion for search and selection of new arenas is consistency with the market/need areas now served by the firm.

An Alternative Approach

Levitt's argument is convincing, but it is not without its detractors. Ansoff, for example, points out that in pursuing a market path in the identification and selection of new business opportunities, there is little or no guarantee of synergy.[14] Synergy — the essential relationship between the new business and the old — is the key variable in the choice of new business areas. Through synergy, Ansoff argues, the whole becomes greater than the sum of its parts; the new business builds on the old, and the combined return on the firm's resources is greater than if each business had been undertaken separately. To extend Levitt's argument, it would have been quite reasonable for the railroads to be in the taxi and car-rental businesses, or even the manufacture of automobiles and trucks. But in those areas, synergy would be lacking. Peters and Waterman, in *In Search of Excellence*, also argue that firms should stick to the knitting — another plea for seeking high-synergy new business areas.[15]

Ansoff makes two major points. First, new business areas should be defined in terms of *product-market* opportunities. Second, synergy should be the key criterion in the identification and selection of new areas for exploitation. This use of both product and market dimensions to characterize opportunity areas goes beyond Levitt's single-market vector approach. Ansoff's view has gained widespread appeal. For example, Corey proposed that two-dimensional matrices, with the dimensions labeled "products" and "markets," be used to identify new business arenas.[16] He notes that markets, together with products that can be developed in response to needs in those markets, define the opportunities for exploitation — the arenas.

Abell takes this matrix approach one step further in *Defining the Business*.[17] He proposes that a business arena be defined in terms of three dimensions:

- *Customer groups served.* For a computer manufacturer, for example, customer groups might include banks, hospitals, manufacturers, and government.
- *Customer functions served.* These might include applications support, services, software, central processing, core memory storage, and disk storage.
- Technologies utilized. For core memory storage, several existing and new technologies might have application to the random access memory function.

Finally, Crawford's study of firms' innovation charters[18] points to several ways in which managers of firms define the arenas:

- by product type;
- by end-user activity;
- by type of technology employed; and
- by end-user group.

Taken individually, each of these ways of defining a business arena has its drawbacks. For example, a product definition of new arenas is limited; products become obsolete, as the railroads found out. Similarly, a market definition could lead the firm into a number of different and unrelated development and production technologies. A technological definition takes the company into all types of different markets, with the many problems that this entails.

How, then, does one define a new business or new-product arena? A combination and adaptation of the approaches suggested above probably works best.[19] Simply stated, a product arena can be described as some combination of

- *who* (the customer group served);
- *what* (the application or customer need); and
- *how* (the technology required to develop and produce the product).

These three dimensions provide a useful starting point for a description of new-product arenas.

Identifying Possible Arenas

The next step is the search for new-product arenas. A three-dimensional diagram is used to illustrate this search (Exhibit 9.4).

Exhibit 9.4.   Defining New-Product Arenas: Dimensions

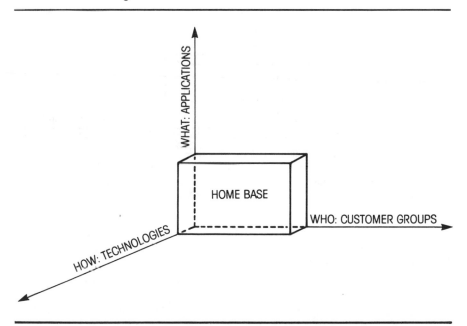

The three dimensions of customer groups, applications, and technologies are shown as the X, Y, and Z axes of the diagram. Locate your home base, and then move away from it on each of the three axes, identifying other customer groups, other applications, and other technologies. In the process, you'll discover a number of new arenas.

An example: A company we'll call Chempro is a medium-sized manufacturer of blending and agitation process equipment for the pulp and paper industry. The company's major strength was its ability to design and manufacture rotary hydraulic equipment. The market served was the pulp and paper industry. The application was agitation and blending of liquids and slurries. The company's current or home base is shown as the shaded cube in Exhibit 9.5.

Exhibit 9.5.   The Arena Dimensions for Chempro

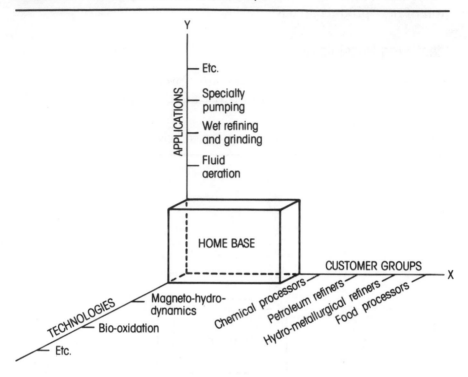

What new-product arenas exist for the company? Clearly, the home base is one of these, and indeed the firm was active in seeking new-product ideas for agitation equipment in the pulp and paper field. Most of these opportunities, however, were limited to modifications and upgrades.

One direction the firm could take is the development of alternative customer groups. These might include the chemical, food-processing, petroleum-refining, and hydrometallurgical fields. The options are shown on the X or horizontal axis of Exhibit 9.5.

Similarly, related applications can be defined. These include the pumping of fluids, fluid aeration, and refining and grinding, as shown on the vertical or Y axis of the arena matrix.

Exhibit 9.6.   Identification of Product Innovation Arenas for Chempro

	CURRENT CUSTOMER GROUP	NEW CUSTOMER GROUPS		
	Pulp and paper	Chemical process industry	Petroleum refining	Hydro metallurigical
CURRENT APPLICATION				
Agitation and blending liquids	Agitators and blenders for pulp and paper industry	Chemical mixers	Petroleum storage blenders	Hydro-metallurgical agitators
NEW APPLICATIONS				
Aeration	Surface aerators for pulp and paper waste treatment lagoons	Aerators for chemical wastes	Aerators for petroleum waste treatment	Aerators for flotation cells (hydro-metallurgy)
Wet refining and grinding	Pulpers, repulpers and refiners			
Specialty pumping	High-density paper stock pumps	Specialty chemical pumps	Specialty petroleum pumps	Slurry pumps

Reprinted with permission from R.G. Cooper "Strategic Planning for Successful Technological Innovation," *Business Quarterly* 43 (Spring 1978): 46–54.

Considering the two dimensions — different applications and different customer groups — we can now proceed to define a number of new arenas. Working with a two-dimensional diagram (Exhibit 9.6), we see that besides the home-base arena there are 12 other arenas the firm could consider for its new-product focus. For example, the firm could develop blending and agitation equipment (same application) aimed at the chemical or petroleum industries (new customer groups). Alternatively, Chempro could target aeration devices (new application) at its current customers, namely pulp and paper firms. Each of these possibilities represents a new arena for the firm.

The firm might also be able to change its third dimension by moving from its home base of rotary hydraulic technology to other technologies. If the alternatives are superimposed along the third dimension atop the matrix, the result is a much larger number of possible arenas. (This third-dimension expansion is not shown in Exhibit 9.6 — it's a little hard on the eyes! Possible alternative arenas include magneto-hydrodynamic pumps and agitators for a variety of end-user groups, bio-oxidation reactors for the food industry, and many others.)

Suggestion: Draw an arena diagram for your company or division. (You may need several diagrams.) Use the three dimensions of customer groups, applications, and technologies. Locate your home base, and then move out on each of the three axes, identifying other customer groups, applications, and technologies. This exercise should help you uncover a number of new but related product arenas.

Selecting the Right Arenas

Now the task is to narrow down the many possible arenas to the target set — the ones that will become the focus of the firm's PIC. To a certain extent, a pre-screening of these arenas has already occurred: each has been identified as being related to the base business on at least one of the three dimensions. Some synergy exists for each arena.

The Criteria

The choice of the right arenas boils down to a single "must" criterion and two "desirable" criteria. The must criterion is an obvious one: Does the arena fit within the corporate mission? The assumptions are that someone has defined a corporate mission and that the current task is the delineation of new-product arenas in light of that mission statement. To the extent that each of the arenas defined in a matrix such as that in Exhibit 9.6 is related to the base business, chances are that most will "pass" this checkpoint. In the Chempro example, all of the arenas in Exhibit 9.6 did pass: each was fair game.

The other two criteria are market attractiveness and business position. Here we use the arena assessment diagram shown in Exhibit 9.7, on which each of the arenas defined in Exhibit 9.6 is replotted in terms of its market attractiveness and the firm's business position.

Market attractiveness is a global dimension that captures how attractive the external opportunity is for an arena. Market attractiveness is shown as the vertical dimension in Exhibit 9.7. Typically, market attractiveness consists of an index score made up of ratings on each of the following questions:

- Is the market in the arena a large one?
- Is it growing?

Exhibit 9.7.   Arena Assessment: Market Attractiveness Versus Business Position

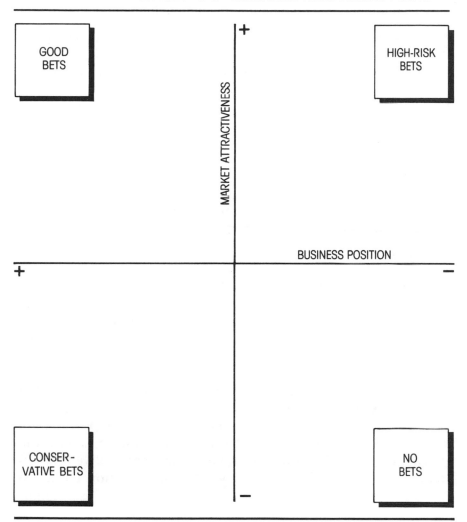

- Does it have good long-term potential?
- Is it over-competitive?
- Is it well served now by competent suppliers?
- Is there a dominant competitor?
- Will it be easy for other firms to enter the arena?

Arenas that feature large or growing markets, good long-term potential, and little competition are the ones that score high on the market attractiveness dimension.

Business position is the horizontal global dimension shown in Exhibit 9.7. Business position focuses on the firm's ability to exploit successfully the defined arena. It is a composite dimension, consisting of answers to the following types of questions:

- In this arena, is there technological synergy? Can the company build on its technological strengths?
- Is there production synergy? Does the arena fit with the firm's production strengths and facilities?
- Is there marketing synergy? Can the company make use of its existing sales force, distribution channels, and customer relationships?
- Does the arena build on the firm's distinctive skills, talents, and resources?
- Can the company use its distinctive competences in order to gain a differential advantage — for example, a strong product advantage — in this arena?

Good Bets

The arena assessment diagram in Exhibit 9.7 is divided into four quadrants. Each quadrant represents a different type of "betting opportunity." The arenas shown in the upper left quadrant, which feature high market attractiveness and a strong business position, are clearly the most desirable. These are called the "good bets." Diagonally opposite, in the lower right quadrant, are the "low-low" arenas — those arenas that neither build on the firm's strengths nor offer attractive external opportunities. These are the "no bets." The "high-risk bets" are in the upper right quadrant: they represent high opportunity arenas in which the firm has a weak business position. Finally, the lower left quadrant houses the "conservative bets" — arenas in which the firm has a good business position but a less attractive market opportunity; these are opportunities that might be exploited at little risk, but will likely yield limited returns.

Assessing the Arenas at Chempro

At Chempro, the problem of arena assessment was simplified by recognizing the firm's technological and financial resource limitations. The company's main asset was its ability to design and engineer rotary hydraulic equipment. The prospect of embarking on new and expensive technologies, such as magneto-hydrodynamics or bio-oxidation, was out of the question. Moreover, having identified its current technology as a field of particular strength, and recognizing that there were many opportunities that could build on this strength, management elected to stay with its current technology. They chose

Exhibit 9.8. Arena Assessment for Chempro

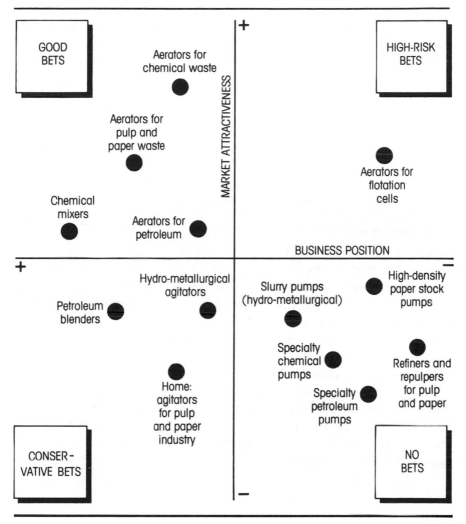

to attack from a position of strength, and so the third dimension — alternative technologies — was erased. The result was the two-dimensional array shown in Exhibit 9.6.

Next, the 12 new arenas plus the home base had to be rated on the two key dimensions of market attractiveness and business position. For this example, a list of questions was employed similar to that set out above; each arena was rated on each question. The questions were then weighted, and a business position and market attractiveness index were computed for each of the 13 possible arenas. The results for Chempro are shown in Exhibit 9.8.

Suggestion: Now that you've identified a list of possible arenas, try to rate each on the two key dimensions of market attractiveness and business position. You might consider developing a list of questions for each dimension, and score each arena. Draw an arena-assessment map (similar to that in Exhibit 9.8) to see where your arenas lie.

Selecting Chempro's Arenas

The choice of arenas depends on the risk/return values of management. Selecting only those arenas in the top half of the arena-assessment diagram — the good bets and the high-risk bets — emphasizes the external magnitude of the opportunity, and places no weight at all on the business-position dimension: a high-return, high-risk choice. The other extreme — selecting only those arenas on the left of the vertical, the good bets and conservative bets — boils down to a low-risk, low-return strategy: select only those arenas in which the firm possesses strong synergy and a good business position. Ideally, one looks for a combination of the two: arenas in which the market attractiveness and the business position both are rated high — the good bets in the upper left quadrant of Exhibit 9.8.

For Chempro, six arenas were rated positively on both dimensions. In order to quantify or rank-order these opportunities, a cutoff line was drawn (see Exhibit 9.9). Arenas to the left of and above this line were considered positive; those to the right and below were negative. The perpendicular distance of each arena from that line was measured. The longer the distance, the more desirable the arena. Based on this exercise, three good bets and one conservative bet were defined as new arenas for Chempro:

- aerators for the chemical industry (waste treatment);
- blenders for the petroleum industry;
- agitators and mixers for the chemical industry;
- surface aerators for the pulp and paper industry.

The decision was made to continue seeking new products in the home-base arena as well. Several other arenas were put on hold for future action.

Suggestion: Use an arena-assessment diagram similar to that shown in Exhibit 9.8 to identify your top-priority arenas. Start with a 45-degree cut-off line, as illustrated in Exhibit 9.9, and measure the distance to each arena. Use a sharper angle for your cut-off line if you want to isolate less risky, lower-return arenas. (Note that a 90-degree

Exhibit 9.9. Assigning Priorities to the Arenas

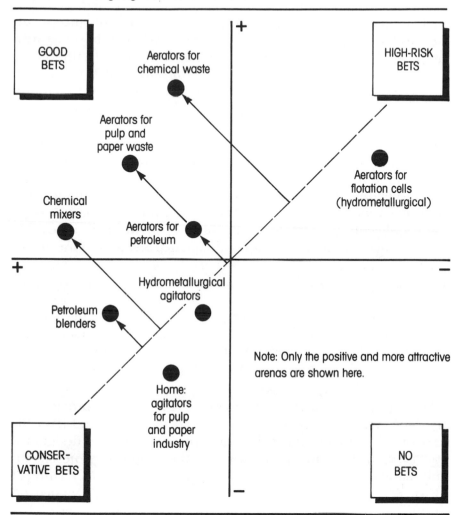

Note: Only the positive and more attractive arenas are shown here.

or vertical line yields only good bets and conservative bets.) A more nearly horizontal cut-off line results in a higher proportion of arenas in the high-risk quadrant.

Putting the Product Innovation Charter to Work

The objectives and the top-priority arenas for the PIC have been defined. Here's how the PIC guides the management of the company's innovation efforts:

Searching for Product Ideas

The definition of objectives and arenas provides guidance to the idea-search effort. Armed with a knowledge of the new business areas the firm wishes to pursue, those in the company charged with seeking new-product ideas will have a clear definition of where to search. Moreover, it now becomes feasible to implement formal search programs — suggestion schemes, contests, sales force call reports, creativity methods, and all the other methods outlined in chapter 4 — to flush out new-product ideas. The search for ideas will be more efficient, generating product ideas that are consistent with the firm's focus.

In Chempro's case, all personnel, from the president to sales trainees, gained a clear view of which new-product arenas the company wished to concentrate on. These new insights made it possible for good new-product ideas in the designated arenas to pour in.

More Effective Screening

One of the most critical screening questions highlighted in chapter 5 was whether the new-product idea fits within the firm's mandate for new products. Put another way, does the firm have any "right" to even consider the product idea? All too often the question is answered with blank stares and shrugs.

A clear delineation of the firm's new-product arenas provides the criterion essential to answer the "mandate fit" question. Either the new-product idea under consideration fits into one of the designated arenas, or it does not. If it does not, the proposed project is an automatic kill. The result is a more effective and efficient screening of ideas. Precious evaluation time and resources are not wasted on product ideas that may seem attractive on their own merits, but simply do not fit into the long-term innovation strategy of the firm.

Manpower and Resource Planning

Resources essential to new products — R & D, engineering, marketing, production — cannot be acquired overnight. Without a definition of which arenas the firm intends to target, planning for the acquisition of these resources is like asking a blindfolded person to throw darts.

For Chempro, aerators for the pulp and paper industry were

defined as a top-priority arena. Research and development management hired researchers in the field of biochemistry and waste treatment; the engineering department acquired new people in the field of aeration design and aeration application engineering, and plans were made to add aeration experts to the sales force. Finally, several exploratory technical and market-research programs were initiated in aeration and bio-oxidation.

The Product Innovation Charter: Conclusion

A PIC is a must for all firms that are serious about building new products into their long-range plans. Many firms operate without such a charter, and the managers in those firms know the problems only too well. There is no direction to the idea search, or there is no idea search at all. Much time is wasted in screening proposed projects and agonizing over the same question: Should we be in this business? Manpower and resource planning is hit and miss, and there are difficulties in securing a long-term, sustained budget commitment for new products from senior management.

A method of defining the PIC has been outlined in this chapter. It begins with a recognition of the need for and rewards of such a charter. Objectives are defined that give the new-product program a sense of purpose, and tie it firmly to the corporation's overall objectives. Arenas — the who, what, and how — are identified and pared down to a set of top-priority fields for exploitation. These arenas give the program direction and focus, ingredients that are critical to a successful innovation strategy.

The War Plan and the Battle Plan

Define the objectives of the new-product program. Choose the arenas on which to focus the program. These are the ingredients of the war plan, or the PIC. Then move toward the battle plan, a step-by-step sequence of actions to bring products from mere ideas to money-making successes in the marketplace. This is the new-product game plan.

There are no guarantees, of course. Even the best war plans and battle plans have been laid waste by bad luck, unforeseen events, and poor execution. Having no plan at all, however, is simply begging disaster to strike. Product innovation is too important to be left to chance and to ad hoc, spur-of-the-moment decision making.

The NewProd
Screening Model

The NewProd screening model is typical of many scoring models used for project evaluation at the early stages of the process.[1] The model consists of a set of questions about the proposed project. The questions serve to characterize the product and its advantages, the attractiveness of its market, the company-project fit, etc.

Up to 12 evaluators are briefed on the project. Independently of one another they answer the questionnaire, providing ratings of the project on each of 48 questions. (The questionnaire and instructions are reproduced in this appendix.) Evaluators also indicate how confident they are about each of their answers.

From the evaluators' inputs, a profile of the project is computed and then reduced to profiles on each of the eight key characteristics that have been found to decide success or failure. These characteristics include:

- product advantage;
- economic advantage of the product to the user;
- corporate synergy (overall fit with the company);
- technological synergy;
- project newness to the firm;
- market need, size, and growth;
- market competitiveness;
- project scope.

From the profile on these eight key factors, the strengths and weaknesses of the project are identified, along with the overall likelihood of commercial success. A sensitivity analysis is also undertaken. (The computations are handled by a microcomputer.)

The major difference between NewProd and other checklist or scoring models is that NewProd is empirically based. That is, both the questions used and the weights on the questions are based on history — on what happened to a large number of past projects. The model predicts success or failure with an accuracy of 84 per cent.

NEWPROD SCREENING MODEL RATING FORM

Section to be filled out by NewProd Model administrator:

Evaluator _____ of _____ evaluators

Project Identification Number _____

Project Name _____

INSTRUCTIONS

- Please read instructions carefully before beginning.
- The NewProd Screening Model is a decision model that combines subjective opinion from a number of evaluators to yield an overall rating on a proposed new-product project.
- As one of these evaluators, you are asked to provide your thoughts or ratings on a number of characteristics of the project identified above (see Project Name, above).
- Please read carefully each of the statements listed in the following pages. Do these characteristics describe the project? Indicate your degree of agreement or disagreement by circling a number from zero (0) to ten (10) on the scale to the immediate right of each statement. Here:
  - 0 means strongly disagree
  - 10 means strongly agree
  - and numbers between 0 and 10 indicate various degrees of agreement or disagreement.
- You must provide a rating for every statement even though you may not be certain about your answer.
- You are also required to indicate how certain or confident you are about each of your responses. Do this by writing a number from 0 to 10 in the column to the far right under the heading ''Confidence.''
  Here:
  - 0 means very low confidence in answer; highly uncertain
  - 10 means total confidence; highly certain
  - and numbers between 0 and 10 indicate varying degrees of confidence.
- So **two** answers are required for each statement:
  - — first, your agreement or disagreement on the 0 to 10 scale;
  - — second, how confident you are in your response, 0 to 10 rating in the far right column.

Example:

	STRONGLY DISAGREE	STRONGLY AGREE	CONFIDENCE (0 TO 10)
1. The product is a totally innovative one.	0 1 ②3 4 5 6 7 8 9 10		_9_

If the product was not innovative, circle a low number on the scale . . . say 2, as noted above. By writing a 9 in the confidence column, you show that you're pretty sure of your assessment of the lack of innovativeness of the product.

Let's begin with some statements about the resource compatibility of this project and our company...

RESOURCES REQUIRED	STRONGLY DISAGREE	STRONGLY AGREE	CONFIDENCE (0 TO 10)
1. Our company's financial resources are more than adequate for this project.	0 1 2 3 4 5 6 7 8 9 10		_____
2. Our company's R & D skills and people are more than adequate for this project.	0 1 2 3 4 5 6 7 8 9 10		_____
3. Our company's engineering skills and people are more than adequate for this project.	0 1 2 3 4 5 6 7 8 9 10		_____
4. Our company's marketing research skills and people are more than adequate for this project.	0 1 2 3 4 5 6 7 8 9 10		_____
5. Our company's management skills are more than adequate for this project.	0 1 2 3 4 5 6 7 8 9 10		_____
6. Our company's production resources or skills are more than adequate for this project.	0 1 2 3 4 5 6 7 8 9 10		_____
7. Our company's sales force and/or distribution resources and skills are more than adequate for this project.	0 1 2 3 4 5 6 7 8 9 10		_____
8. Our company's advertising and promotion skills and resources are more than adequate for this project.	0 1 2 3 4 5 6 7 8 9 10		_____

NATURE OF PROJECT	STRONGLY DISAGREE	STRONGLY AGREE	CONFIDENCE (0 TO 10)
9. Our product is highly innovative — totally new to the market.	0 1 2 3 4 5 6 7 8 9 10		_____
10. Our product is a very high-technology one.	0 1 2 3 4 5 6 7 8 9 10		_____
11. Our product is a "big ticket item" — it will sell for a very high per-unit price.	0 1 2 3 4 5 6 7 8 9 10		_____

	STRONGLY DISAGREE	STRONGLY AGREE	CONFIDENCE (0 TO 10)
12. Our product is mechanically and/or technically very complex.	0 1 2 3 4 5 6 7 8 9 10		_____
13. The product **idea** came to us from the marketplace, as opposed to in-house lab or technical work (10 = marketplace; 0 = in-house).	0 1 2 3 4 5 6 7 8 9 10		_____
14. The product specifications — exactly what the product should be — are very clear from the beginning of the project.	0 1 2 3 4 5 6 7 8 9 10		_____

15. The technical aspects — exactly how the technical problems would be solved — are very clear from the beginning.

0 1 2 3 4 5 6 7 8 9 10   _____

16. Our product is a custom product — designed for each customer — as opposed to a standard product (10 = custom; 0 = standard).

0 1 2 3 4 5 6 7 8 9 10   _____

17. Our product is a defensive introduction to maintain our market share in the market as opposed to gaining share or new customers (10 = defensive; 0 = offensive).

0 1 2 3 4 5 6 7 8 9 10   _____

18. Relative to our other new product introductions, the expenses and investment incurred up to the first sale of the product will be considerably greater (10 = considerably greater; 0 = considerably less).

0 1 2 3 4 5 6 7 8 9 10   _____

NEWNESS OF PROJECT

STRONGLY DISAGREE   STRONGLY AGREE   CONFIDENCE (0 TO 10)

19. The potential customers for this product are totally new to our company.

0 1 2 3 4 5 6 7 8 9 10   _____

20. The product class itself is totally new to our company.

0 1 2 3 4 5 6 7 8 9 10   _____

21. We have never made or sold products to satisfy this type of customer need or use before.

0 1 2 3 4 5 6 7 8 9 10   _____

22. The nature of the production process is totally new to our company.

0 1 2 3 4 5 6 7 8 9 10   _____

23. The technology required to develop the product (R & D) is totally new to our company.

0 1 2 3 4 5 6 7 8 9 10   _____

24. The distribution system and/or type of sales force for this product are totally new to our company.

0 1 2 3 4 5 6 7 8 9 10   _____

25. The type of advertising and promotion required is totally new to our company.

0 1 2 3 4 5 6 7 8 9 10   _____

26. The competitors we face in the market for this product are totally new to our company.

0 1 2 3 4 5 6 7 8 9 10   _____

THE FINAL PRODUCT

STRONGLY DISAGREE   STRONGLY AGREE   CONFIDENCE (0 TO 10)

27. Compared to competitive products, our product will offer a number of unique features or attributes to the customer.

0 1 2 3 4 5 6 7 8 9 10   _____

28. Our product will be clearly superior to competing products in terms of meeting customers' needs.

0 1 2 3 4 5 6 7 8 9 10 _____

29. Our product will permit the customer to reduce his costs when compared to what he is now using.

0 1 2 3 4 5 6 7 8 9 10 _____

30. Our product will permit the customer to do a job or do something he cannot presently do with what is now available.

0 1 2 3 4 5 6 7 8 9 10 _____

31. Our product will be of higher quality — tighter specifications or stronger or will last longer or be more reliable, etc. — than competing products.

0 1 2 3 4 5 6 7 8 9 10 _____

32. Our product will be priced considerably higher than competing products (10 = much higher; 0 = much lower).

0 1 2 3 4 5 6 7 8 9 10 _____

33. We will be the first into the market with this type of product.

0 1 2 3 4 5 6 7 8 9 10 _____

OUR MARKET FOR THIS PRODUCT

STRONGLY DISAGREE   STRONGLY AGREE   CONFIDENCE (0 TO 10)

34. There are many potential customers for this product — a mass market — as opposed to one or a few customers (10 = mass market; 0 = one customer).

0 1 2 3 4 5 6 7 8 9 10 _____

35. Potential customers have a great need for this class of product.

0 1 2 3 4 5 6 7 8 9 10 _____

36. There is only a "potential demand" for this product class — no market exists currently.

0 1 2 3 4 5 6 7 8 9 10 _____

37. The dollar size of the market (either existing or potential) for this product is large.

0 1 2 3 4 5 6 7 8 9 10 _____

38. The market for this product is growing very quickly.

0 1 2 3 4 5 6 7 8 9 10 _____

39. Competing products (or whatever the customer is now using) are very similar to each other — a high degree of product homogeneity.

0 1 2 3 4 5 6 7 8 9 10 _____

40. The market is a highly competitive one.
0 1 2 3 4 5 6 7 8 9 10 _____

41. The market is characterized by intense price competition.

0 1 2 3 4 5 6 7 8 9 10 _____

42. There are many competitors in this market.

0 1 2 3 4 5 6 7 8 9 10 _____

43. There is a strong, dominant competitor — with a large market share — in the market.

0 1 2 3 4 5 6 7 8 9 10 _____

44. There is a high degree of loyalty to existing (competitors') products in this market.   0 1 2 3 4 5 6 7 8 9 10 _____

45. Potential customers are very satisfied with the products they are currently using (competitors' products).   0 1 2 3 4 5 6 7 8 9 10 _____

46. New product introductions by competitors are frequent in this market.   0 1 2 3 4 5 6 7 8 9 10 _____

47. Users' needs change quickly in this market — a dynamic market situation.   0 1 2 3 4 5 6 7 8 9 10 _____

48. Government legislation, rules, standards, etc. play an important role in the design and testing of products for this market.   0 1 2 3 4 5 6 7 8 9 10 _____

Please check over your responses to all the questions to make sure you have answered every question.

Your Name _____

Signature _____

Date _____

Phone _____

Benefit-Contribution Screening Methods: Financial Indices

A number of different financial indices have been proposed over the years to yield a short-cut financial assessment of a proposed new-product project.[1] The simplest are straightforward ratios; for example,

$$\text{Index} = \frac{S \cdot \sqrt{L}}{C} \tag{1}$$

where S is the expected sales of the product in a typical year, L is the product's economic life on the market, and C is the total cost to implement the product (development and commercialization). (All variables are defined at the end of this appendix.) As with most indices, the appropriate or cut-off value — the value of the index needed to signal a GO decision — is arbitrarily determined and company-specific.

A second ratio, called the cost ratio, brings in the probability of technical success of the project, P_t. But only R & D costs are considered. The index is:

$$\text{Cost ratio} = \frac{P_t \cdot Pr}{RD} \tag{2}$$

The use of profits, Pr, or income for a typical year, rather than sales, requires considerably more data than does index number 1.

Pacifico proposes yet another index, and one similar to the cost ratio.[2] He includes both the probability of technical success, P_t, and the probability of commercial success, given a technical success, P_c. Additionally, both investment and R & D costs are included:

$$\text{Index} = \frac{P_t \cdot P_c \cdot L \cdot Pr}{I + RD} \tag{3}$$

The square root of the product's economic life has also been used in this formula in order to discount future years.

One of the most complicated formulas is a combination one suggested by Teal. The four sub-indices included are R & D index; investment index; contribution to sales index; and market share index. Teal's index is:

$$
\text{Index} = \left(\frac{\sum\limits^{L} \text{Pr}}{25 \, \text{RD}} \right) \cdot \left(\frac{\sum\limits^{L} \text{Pr}}{.27 \, (\text{I} + \text{RD})} \right) \cdot \left(\frac{25\text{S}}{\text{T}} \right) \cdot \left(\frac{2\text{S}}{\text{MS}} \right) \quad (4)
$$

where T is the total company sales and MS is the total market size.

One index, proposed by Disman, introduces the notion of discounted profits. Here R is the cost of capital, expressed as a decimal (for example, 0.15):

$$
\text{Index} = \frac{P_t \cdot P_c \cdot 2 \sum\limits^{L} \dfrac{\text{Pr}}{(\text{I} + \text{R})^n}}{\text{RD}} \quad (5)
$$

Finally, another index is based on a discounted return method.[3] The return, expressed as a percentage is:

$$
i = \frac{P_s \, (\text{S} \cdot \text{M} - \text{CE})}{P_s \, (\text{D} + \text{LC} + \text{I}) + (\text{I} - P_s) \, \text{F}} \cdot 100\% \quad (6)
$$

where D is the development cost, LC is the launch cost, I is the investment, and F is the loss that would be incurred in the event of failure. For the special case where F = D + LC + I, the return reduces to:

$$
i = \frac{P_s \, (\text{S} \cdot \text{M} - \text{CE})}{\text{D} + \text{LC} + \text{I}} \cdot 100\% \quad (7)
$$

The derivation of this last index is as follows. A new-product project is developed in year 0, incurring a development cost D. The total investment to commercialize the product (new plant, tooling, working capital, etc.) is I. The launch or marketing costs are LC. Thus, the total cash disbursements to develop and commercialize the product are D + LC + I.

The product is launched, achieves an annual sales level of S, a profit margin rate of M (expressed as a decimal), and annual cash operating expenses (excluding depreciation) of CE. These are figures for a typical year and remain constant for the economic life of the product, L, which is very long. The cash profit (cash inflow minus outflow) for any year is thus $S \cdot M - CE$. The present value of these profits over the life of the product is:

$$PV = \frac{S \cdot M - CE}{1 + i} + \frac{S \cdot M - CE}{(1 + i)^2} + \frac{S \cdot M - CE}{(1 + i)^3} + \quad \cdots$$

Here i is the discount rate. If L is very long, then this is an infinite series whose first term is $(S \cdot M - CE)/(1 + i)$ and whose ratio is $1/(1 + i)$. The sum of such a series is:

$$PV = \frac{S \cdot M - CE}{i}$$

If the product is a commercial and technical success, then the consequences or net present value are simply the discounted future cash profits less the initial cash outlays:

$$NPV = \frac{S \cdot M - CE}{i} - (D + LC + I)$$

If the product should fail, let F be the cost incurred — the cost of failure. Note that F will likely be less than the initial cash disbursements, since some of these may be recoverable in the event of a failure. So, if a failure:

$$NPV = - F$$

Let P_s be the probability of technical and commercial success. The expected net present value is:

$$ENPV = P_s \left[\frac{S \cdot M - CE}{i} - (D + LC + I) \right] - (I - P_s) \cdot F$$

To determine the return, let the ENPV = 0 and solve for i, the internal rate of return:

$$P_s (S \cdot M - CE) = P_s \cdot i (D + LC + I) + (I - P_s) \cdot i \cdot F$$

$$i = \frac{P_s (S \cdot M - CE)}{P_s (D + LC + I) + (I - P_s) F} \cdot 100\% \tag{6}$$

and for the special case where $F = D + LC + I$, then:

$$i = \frac{P_s (S \cdot M - CE)}{D + LC + I} \cdot 100\% \qquad (7)$$

Nomenclature:

C = cost to develop and commercialize the product.

CE = cash operating expenses for a year.

$ENPV$ = expected net present value.

F = cost of failure.

I = total investment in the product.

i = internal rate of return or discount rate (as a decimal).

L = economic life of the product (years).

M = profit margin, as a proportion (decimal) of sales.

MS = market size.

NPV = net present value.

Pr = the annual profits or income from the product in a typical year.

PV = present value.

P_c = probability of commercial success, given a technical success.

P_s = overall probability of success (both technical and commercial).

P_t = probability of technical success.

R = cost of capital (as a decimal).

RD = the R & D or development costs of the product.

S = annual sales of the product for a typical year.

T = total company sales.

Survey Methods

Telephone Surveys

Telephone surveys are relatively inexpensive, fast to execute, and simple to design. They yield a good response rate, particularly if the respondent has an interest in the product. They do have their limitations, however.

First, a telephone listing of the sample must be available, a listing that is representative of the target market. Second, the information that can be collected by phone is limited because the duration of the interview is necessarily short — only a handful of questions can be asked. Third, the types of questions used and information obtained are limited. No pictures, visuals, or samples can be used, and scaled questions and questions with multiple response categories are almost impossible to ask over the phone. In the truck study cited in chapter 6, because so much information of a fairly complex nature was sought, a phone survey was ruled out in spite of its economic attractiveness.

Hints

- If possible, start with the names of people, not just organization names or phone numbers. If that is not possible, then build a qualifying section into the interview to ensure that you're talking to the right person. You may also require a more skilled interviewer if qualification is required.
- Use a structured questionnaire. This is a research study, not a chatty telephone conversation. Questions should be worded carefully and clearly. The questionnaire should be pretested on a limited number of people, and response categories should be provided to facilitate the recording or answers by the interviewer.

- Select your interviewers carefully. A dolt of an interviewer can ruin an otherwise excellently designed study. One major packaged-goods firm hires teachers for telephone interviewing at night: teachers are typically literate, well-spoken, intelligent, and relatively at ease in dealing with others. Make sure you train your interviewers — they must understand the contents of the questionnaire and what information is sought.

Mail Surveys

At first glance, mail surveys appear most attractive, but they are also fraught with difficulties. On the plus side, they are ideal for large sample sizes: the variable cost per respondent is little more than the cost of paper and a stamp. But the design and set-up costs are probably the highest of any research method.

The mail survey questionnaire must be letter-perfect. Unlike the other methods, there's no interviewer present to explain what you really meant by that question. Questions must be carefully designed and tested, and operational definitions of terms used must be clearly spelled out. For example, in a mail questionnaire, the apparently simple question, "How many rooms are there in your house?" is loaded with problems. What do you mean by "my house"? I live with my wife's folks, but I own a summer cottage — which house do I count? And what's a "room" — do I count bathrooms, finished basements, bedrooms in attics, garages, Florida rooms? Depending on how the respondent reads the question, he could answer anywhere from 5 to 15 rooms — an unreliable response.

Besides the attractive economics of large sample sizes, mail surveys also have the advantage of being able to collect more information from more questions than is possible in the typical phone survey. Moreover, the questions and information sought can be more flexible: visuals, pictures, and even samples can be shown — for example, a sample of cloth, or two or three line drawings of a proposed product. Multiple response categories and scaled questions can also be used in mailed questionnaires.

There are three major difficulties with a mail survey. The first is that there is no interviewer present to explain the questions and to interpret and record the responses. The questionnaire must be perfectly worded, the questions unambiguous and easy to answer, and operational definitions of terms used must be given. Probing questions, which often lead to the richest information, are hard to handle in a mail survey.

The second obstacle is the low response rate: it's much easier to toss a mailed questionnaire into the wastebasket than it is to hang up the phone on someone or to throw someone out of an office. And even if the response rate is a respectable 20 to 30 per cent, you still face a dilemma. You must generalize from the questionnaires that are returned. The assumption is that the people who didn't reply would have answered the same way! That assumption is often wrong. In short, you may have a very biased sample of respondents.

The third problem is getting the right person to answer the questionnaire. There's little chance to qualify the respondent by mail; if it's sent to a household, who answers it? The problem is worse when you send the questionnaire to a company, unless you've prequalified the person by telephone, for example. Getting your hands on the appropriate mailing list, with names and not just addresses, may be difficult for some product types.

Although a mail survey is intuitively appealing, be very cautious with this method. If you have little experience with a mail survey, hire a professional to design the study.

Hints

- Watch the wording of questions: avoid ambiguity, and provide operational definitions of terms such as "your house" and "rooms."
- No matter how careful you are in wording the questionnaire, always pretest it on others — first your colleagues, then a limited number of respondents. Follow up the pretest with a visit or phone call to go over the questionnaire.
- Make the questionnaire as easy to answer and return as possible. Provide a stamped and addressed return envelope. Don't ask questions that are likely to prove frustrating, threatening, or embarrassing to answer. Provide response categories requiring only a checkmark, not a written answer. Guarantee anonymity.
- Qualify the mailing list, and get names of people, not just names of companies or household addresses. If necessary, do some preliminary work by telephone to qualify and sharpen the list, and to ensure that the right person receives the questionnaire.
- Strive to make the response rate as high as possible to ensure representativeness. Make the mailing a professional-looking piece — typeset and printed, perhaps on colored paper. Provide an incentive, such as a small gift either enclosed with the mailing or awarded when the questionnaire is returned.

- To handle the possible response bias in the event of a low response rate, consider doing a second mailing or second wave to the nonrespondents. Precode these questionnaires differently from the first wave, perhaps by using a different color paper. Compare the responses from the second wave to the first wave. If they're the same, you can tentatively conclude that the initial nonrespondents would have answered the same way as the initial respondents — that the first-wave repondents are indeed representative of the whole sample. Alternatively, measure characteristics of the respondent: age, sex, location, etc., or, in the case of firms, size, industry, location, etc. Compare the demographics of the respondents to the demographics of the original sample. If they are widely different, you may have a biased set of returns. In your data analysis, you can weight your returned questionnaires differently to adjust for this bias.

Personal Interviews

Personal interviews are the most expensive but probably the most effective of the three methods, particularly for pre-development market studies in a new-product project. For an industrial-product study, the cost of an interviewer is as high as or higher than the cost of a sales call. Good interviewers, particularly in the case of engineered or technically complex products, are hard to find and expensive. Travel costs can also mount up. Consumer-goods interviews are generally less expensive. The interviewer is usually not a professional and costs less per interview, and interviews can often be done in one location, such as in a shopping center or at a ski resort (with permission), thereby minimizing interviewer travel time and costs. (Incidentally, the one-stop interview can be used for industrial goods. I've used trade shows and even truck stops as locations for multiple personal interviews.)

The main advantages of personal interviews are the types and the amount of information that can be gathered. There is great flexibility in the types of questions that can be asked: closed-ended questions using multiple response categories or scales; open-ended questions requiring probing; questions involving pictures, drawings, and even samples. The interviewer is present to assist the respondent in replying and to record the answers. (Of course, interviewer bias is always a problem, and interviewer training is essential.) Such interviews are also typically longer than mail or phone, and the topics covered and number of questions can be substantially greater than for either of those methods. Finally, the response rates are higher than for mail, and the respondent can be qualified.

Hints

- Use a questionnaire. Questions should be carefully and clearly worded, avoiding ambiguity and providing operational definitions. Pretest the questionnaire.
- Take advantage of the fact that the interview is face-to-face. Use visuals, pictures, and samples.
- Carefully select and thoroughly train your interviewers. Sit in on the first few interviews yourself.

Notes

Chapter 1
1. Harvard Business School case. Disguised and adapted with permission.
2. *New Product Management for the 1980s* (New York: Booz Allen & Hamilton, 1982).
3. D.S. Hopkins, *New Products Winners and Losers*, report no. 773 (New York: The Conference Board, 1980), 1.
4. Ibid., 3.
5. R.G. Cooper, "New Product Success in Industrial Firms," *Industrial Marketing Management* 11 (1982): 215-23. See also R.G. Cooper, "The Performance Impact of Product Innovation Strategies," *European Journal of Marketing* 18 (1984): 1-54.
6. *New Products Management for the 1980s*, 2-4.
7. Cooper, "New Product Success in Industrial Firms." See also R.G. Cooper, "The Impact of New Product Strategies," *Industrial Marketing Management* 12 (1983): 243-56, R.G. Cooper, "The Performance Impact of Product Innovation Strategies," R.G. Cooper, "How New Product Strategies Impact on Performance," *Journal of Product Innovation Management* 1 (1984): 5-18, R.G. Cooper, "The Strategy-Performance Link in Product Innovation," *R & D Management* 14 (1984): 247-59, R.G. Cooper, "New Product Strategies: What Distinguishes the Top Performers," *Journal of Product Innovation Management* 2 (1984): 151-64, R.G. Cooper, "Industrial Firms' New Product Strategies," *Journal of Business Research* 13 (1985): 107-21, R.G. Cooper, "Overall Corporate Strategies for New Product Programs," *Industrial Marketing Management* 14 (1985): 179-83.
8. *New Product Management for the 1980s*, 5.
9. Ibid., 12-13.

Chapter 2
1. C.M. Crawford, "New Product Failure Rates — Facts and Fallacies," *Research Management* (Sept. 1979): 9-13.
2. R.G. Cooper, "New Product Success in Industrial Firms," *Industrial Marketing Management* 11 (1982): 215-23. See also R.G. Cooper, "The Performance Impact of Product Innovation Strategies," *European*

Journal of Marketing 18 (1984): 1-54, and R.G. Cooper, "Overall Corporate Strategies for New Product Programs," *Industrial Marketing Management* 14 (1985): 179-83.

3. D.S. Hopkins, *New Products Winners and Losers*, report no. 773 (New York: The Conference Board, 1980).

4. *New Product Management for the 1980s* (New York: Booz Allen & Hamilton, 1982), 7.

5. Ibid., 14.

6. Hopkins, *New Products Winners and Losers*, 7.

7. R.G. Cooper, "The Impact of New Product Strategies," *Industrial Marketing Management* 12 (1983): 243-56. See also R.G. Cooper, "The Performance Impact of Product Innovation Strategies," *European Journal of Marketing* 18 (1984): 1-54, R.G. Cooper, "How New Product Strategies Impact on Performance," *Journal of Product Innovation Management* (1984): 5-18, R.G. Cooper, "The Strategy-Performance Link in Product Innovation," *R & D Management* 14 (Oct. 1984): 247-59, R.G. Cooper, "New Product Strategies: What Distinguishes the Top Performers," *Journal of Product Innovation Management* 2 (1984): 151-64, R.G. Cooper, "Industrial Firms' New Product Strategies," *Journal of Business Research* 13 (April 1985): 107-21, and R.G. Cooper, "Overall Corporate Strategies for New Product Programs," *Industrial Marketing Management* 14 (1985): 179-83.

8. D.S. Hopkins and E.L. Bailey, "New Product Pressures," *Conference Board Record* 8 (1971): 16-24. See also Hopkins, *New Products Winners and Losers*.

9. Hopkins, *New Products Winners and Losers*, 12-13.

10. R.G. Cooper, "Why New Industrial Products Fail," *Industrial Marketing Management* 4 (1975): 315-26.

11. Ibid. See also Hopkins and Bailey, "New Product Pressures."

12. R. Calantone and R.G. Cooper, "A Discriminant Model for Identifying Scenarios of Industrial New Product Failure," *Journal of the Academy of Marketing Science* 7 (1979): 163-83.

13. S. Myers and D.G. Marquis, *Successful Industrial Innovations* (Washington: National Science Foundation, 1969).

14. R.W. Roberts and J.E. Burke, "Six New Products — What Made Them Successful," *Research Management* 14 (May 1974): 21-24.

15. R.G. Cooper, *Winning the New Product Game* (Montreal: McGill University, 1976). See also R.G. Cooper, "Introducing Successful New Products," *European Journal of Marketing* (1976), MCB monograph series.

16. R. Rothwell, "Factors for Success in Industrial Innovations," in *Project SAPPHO — A Comparative Study of Success and Failure in Industrial Innovation* (Brighton, U.K.: Science Policy Research Unit, University of Sussex, 1976). See also R. Rothwell et al., "SAPPHO Updated — Project SAPPHO Phase II," *Research Policy* 3 (1974): 258-91.

17. R. Rothwell, "The Hungarian SAPPHO: Some Comments and Comparison," *Research Policy* 3 (1974): 30-38.
18. R. Rothwell, "Innovation in Textile Machinery: Some Significant Factors in Success and Failure," *SPRU Occasional Paper Series,* no. 2 (Brighton, U.K.: University of Sussex, 1976).
19. A.H. Rubenstein et al., "Factors Influencing Innovation Success at the Project Level," *Research Management* (May 1976): 15-20.
20. M.A. Maidique and B.J. Zirger, "A Study of Success and Failure in Product Innovation: The Case of the U.S. Electronics Industry," *IEEE Trans. Engineering Management,* EM-31 (Nov. 1984): 192-203.
21. R.G. Cooper, "Identifying Industrial New Product Success: Project NewProd," *Industrial Marketing Management* 8 (1979): 124-35. See also R.G. Cooper, "The Dimensions of Industrial New Product Success and Failure," *Journal of Marketing* 43 (Summer 1979): 93-103, R.G. Cooper, "Project NewProd: Factors in New Product Success," *European Journal of Marketing* 14 (1980): 277-92, R.G. Cooper, *Project NewProd: What Makes a New Product a Winner* (Montreal: Quebec Industrial Innovation Center, 1980), R.G. Cooper, "The Myth of the Better Mousetrap: What Makes a New Product a Success," *Business Quarterly* 46 (Spring 1981): 69-81, R. Calantone and R.G. Cooper, "New Product Scenarios: Prospects for Success," *Journal of Marketing* 45 (Spring 1981): 48-60.
22. *New Product Management for the 1980s,* (New York: Booz Allen & Hamilton, 1982).

Chapter 3

1. R.G. Cooper, *Winning the New Product Game* (Montreal: McGill University, 1976). See also R.G. Cooper, "Introducing Successful New Products," *European Journal of Marketing* (1976), MCB monograph series.
2. R. Rothwell, "The Characteristics of Successful Innovations and Technically Progressive Firms (With Some Comments on Innovation Research)," *R & D Management* 7 (1977): 191-206.
3. Ibid.
4. *New Product Management for the 1980s.* (New York: Booz Allen & Hamilton, 1982). See also More's work on the timing of market-research expenditures: R.A. More, "Timing of Market Research in New Industrial Product Situations," *Journal of Marketing* 48 (Fall 1984): 84-94.
5. Much of this section is taken from R.G. Cooper, "A Process Model for Industrial New Product Development," *IEEE Trans. Engineering Management* EM-30 (Feb. 1983): 2-11.
6. My observations of what happened in over 60 actual case histories laid the foundation for the game plan. See R.G. Cooper, "The New Product Process: An Empirically Derived Classification Scheme," *R & D Management* 13 (Jan. 1983): 1-13. See also R.G. Cooper, *Winning the New*

Product Game (Montreal: McGill University, 1976). and R.G. Cooper, "Introducing Successful New Products," *European Journal of Marketing* (1976), MCB monograph series.

7. R. Rothwell, "Innovation in Textile Machinery: Some Significant Factors in Success and Failure," SPRU Occasional Paper Series, no. 2 (Brighton, U.K.: University of Sussex, 1976).

8. C.M. Crawford, "Protocol: New Tool for Product Innovation," *Journal of Product Innovation Management* 2 (1984): 85-91.

9. Cooper, *Winning the New Product Game.* See also Cooper, "Introducing Successful New Products."

Chapter 4

1. E.A. Von Hippel, "Has Your Customer Already Developed Your Next Product?" *Sloan Management Review* (Winter 1977): 63-74. See also E.A. Von Hippel, "Successful Industrial Products from Customer Ideas," *Journal of Marketing* (Jan. 1978): 39-49, J. Kreiling and E.A. Von Hippel, "Users Dominate Much Instrument Innovation," *Instrument Technology* (Feb. 1979): 7, and E.A. Von Hippel, "Get New Products from Customers," *Harvard Business Review* (Mar.-April 1982): 117-22.

2. Von Hippel, "Has Your Customer Already Developed Your Next Product?" See also Von Hippel, "Get New Products from Customers."

3. C.M. Crawford, "Unsolicited New Product Ideas: Handle with Care," *Research Management* (Jan. 1975): 22.

4. *Generating New Product Ideas*, report no. 546 (New York: The Conference Board, 1972).

5. M.S. Basadur and R. Thompson, "Usefulness of the Ideation Principle of Extended Effort in Real World Professional and Managerial Problem Solving," forthcoming.

6. R.C. Bennett and R.G. Cooper, "The Misuse of Marketing: An American Tragedy," *Business Horizons* (Nov.-Dec. 1981): 51-61.

7. R.C. Bennett and R.G. Cooper, "The Product Life Cycle Trap," *Business Horizons* (Sept.-Oct. 1984): 7-16.

Chapter 5

1. U. de Brentani, "Evaluation of Industrial New Product Ideas: An Empirical Study," PH.D. dissertation, McGill University (1983).

2. A. Albala, "Stage Approach for the Evaluation and Selection of R & D Projects," *IEEE Trans. Engineering Management* EM-22 (Nov. 1975): 153-62.

3. R.G. Cooper, "An Empirically Derived New Product Project Selection Model," *IEEE Trans. Engineering Management* EM-28 (Aug. 1981): 54-61. Much of this chapter is based on this article and on a subsequent monograph: R.G. Cooper, *A Guide to the Evaluation of New Industrial Products for Development* (Montreal: Industrial Innovation Center, 1982).

4. Albala, "Stage Approach for the Evaluation and Selection of R & D Projects."

5. Ibid.

6. N.R. Baker and W. Pound, "R & D Project Selection: Where We Stand," *IEEE Trans. Engineering Management* EM-11 (Dec. 1964): 124-34. See also A. Charles and A.C. Stedry, "Chance Constrained Model for Real-Time Control in Research and Development Management," *Management Science* 12 (April 1966): B353-62, G.R. Glotskey, "Research on a Research Department: An Analysis of Economic Decisions on Projects," *IEEE Trans. Engineering Management* EM-7 (Dec. 1960): 166-72, and J.R. Moore Jr. and N.R. Baker, "Computational Analysis of Scoring Models for R & D Project Selection," *Management Science* 16 (Dec. 1969): B212-32.

7. N.R. Baker, "R & D Project Selection Models: An Assessment," *IEEE Trans. Engineering Management* EM-21 (Nov. 1974): 165-71. See also N.R. Baker and J. Freeland, "Recent Advances in R & D Benefit Measurement and Project Selection Methods," *Management Science* 21 (1975): 1164-75, and E.P. McGuire, *Evaluating New Product Proposals*, report no. 604 (New York: Conference Board, 1973).

8. Baker, "R & D Project Selection Models: An Assessment."

9. Ibid. See also A. Hansson et al., *Om Formella Metodor for Vardering och Val av Forsknings — och Utvecklings Projekt i Svenska Industriforetag — en enkatundersoking* (Stockholm: Stockholm University, Sept. 1971), and L.M. Katz, "New Product Screening Techniques," MBA dissertation, McGill University (1974).

10. Baker and Freeland, "Recent Advances in R & D Benefit Measurement and Project Selection Methods."

11. Katz, "New Product Screening Techniques." See also McGuire, *Evaluating New Product Proposals.*

12. R.A. More and B. Little, "The Application of Discriminant Analysis to the Prediction of Sales Forecast Uncertainty in New Product Situations," *Journal of the Operational Research Society* 31 (1980): 71-77. See also R.A. More, "Sales Forecast Uncertainty in the Screening of New Industrial Products: A Descriptive Model with Predictive Implications," PH.D. dissertation, University of Western Ontario (1975).

13. McGuire, *Evaluating New Product Proposals.*

14. W.E. Souder, "A Scoring Methodology for Assessing the Suitability of Management Science Models," *Management Science* 18 (June 1972): B526-43.

15. W.E. Souder, "A System for Using R & D Project Evaluation Methods," *Research Management* 21 (Sept. 1978): 29-37.

16. D.R. Augood, "Review of R & D Evaluation Methods," *IEEE Trans. Engineering Management* EM-20 (Nov. 1973): 114-20.

17. Baker and Freeland, "Recent Advances in R & D Benefit Measurement."

18. Augood, "Review of R & D Evaluation Methods."

19. R.G. Cooper, "Selecting Winning New Product Projects: Using the NewProd System," *Journal of Product Innovation Management* no. 2 (1985): 34-44.
20. R.G. Cooper, "An Empirically Derived New Product Project Selection Model."
21. Ibid. See also R.G. Cooper, *A Guide to the Evaluation of New Industrial Products,* and R.G. Cooper, "Selecting Winning New Product Projects." "NewProd" is a registered trade name of R.G. Cooper & Associates Consultants Inc.
22. W.E. Souder, "Achieving Organizational Consensus with Respect to R & D Project Selection Criteria," *Management Science* (Feb. 1975): 669-81.

Chapter 6

1. C.M. Crawford, "Protocol: New Tool for Product Innovation," *Journal of Product Innovation Management* 2 (1984): 85-91.
2. G.L. Urban and J.R. Hauser. *Design and Marketing of New Products,* Englewood Cliffs, N.J.: Prentice-Hall, 1980. A more detailed description of these and other market research techniques and related mapping methods can be found in J.M. Choffrey and G. Lillien, *Marketing Planning for New Industrial Products* (New York: John Wiley & Sons, 1980).
3. The illustration is taken from my own case histories. Identifying details have been altered.

Chapter 7

1. R.G. Cooper, *Winning the New Product Game* (Montreal: McGill University, 1976). See also R.G. Cooper, "Introducing Successful New Products," *European Journal of Marketing* (1976), MCB monograph series.
2. Conversation with Dr. D. Ennis of Philip Morris Inc.
3. R.G. Cooper, "Why New Industrial Products Fail," *Industrial Marketing Management* 4 (1975): 315-26.
4. For a detailed "how to do it" approach to financial analysis, see M.D. Rosenau Jr., *Innovation: Managing the Development of Successful New Products* (Belmont, Calif.: Lifetime Learning Publications, 1982), chapter 4.

Chapter 8

1. R.D. Buzzell et al., "Market Share — A Key to Profitability," *Harvard Business Review* (Jan.-Feb. 1975): 97-107. See also S. Schoeffler, "Impact of Strategic Planning on Profit Performance," *Harvard Business Review* (March-April 1974): 137-45.
2. *The Product Portfolio,* pamphlet no. 66 (Boston: The Boston Consulting Group, 1970).

3. R.C. Bennett and R.G. Cooper, "The Misuse of Marketing: An American Tragedy," *Business Horizons* (Nov.-Dec. 1981): 51-61.
4. D. Ogilvy, *Ogilvy on Advertising* (London: Orbis Publishing and Pan Books, 1983).

Chapter 9
1. C.M. Crawford, "Defining the Charter of Product Innovation," *Sloan Management Review* (Fall 1980): 3-12.
2. Parts of this chapter are based on R.G. Cooper, "Strategic Planning for Successful Technological Innovation," *Business Quarterly* 43 (Spring 1978): 46-54.
3. D.J. Luck and A.E. Prell, *Market Strategy* (Englewood Cliffs, N.J.: Prentice-Hall, 1968), at 2. See also D.J. Luck and O.C. Ferrel, *Marketing Strategy and Plans* (Englewood Cliffs, N.J.: Prentice-Hall, 1979), at 6.
4. I.H. Ansoff, *Corporate Strategy* (New York: McGraw-Hill, 1965), at 103.
5. R.E. Corey, "Key Options in Market Selection and Product Planning," *Harvard Business Review* (Sept.-Oct. 1978): 119-28.
6. *New Product Management for the 1980s.* (New York: Booz Allen & Hamilton, 1982).
7. Ibid.
8. R.D. Buzzell et al., "Market Share — A Key to Profitability," *Harvard Business Review* (Jan.-Feb. 1975): 97-107. See also S. Schoeffler, "Impact of Strategic Planning on Profit Performance," *Harvard Business Review* (March-April 1974): 137-45.
9. H. Nystrom, *Creativity and Innovation* (New York: John Wiley & Sons, 1979). See also H. Nystrom, "Company Strategies for Research and Development," in *Industrial Innovation*, edited by N. Baker (New York: Macmillan, 1979), H. Nystrom, *Company Strategies for Research and Development* (Uppsala: Institute for Economics and Statistics, 1978), H. Nystrom and B. Edvardsson, *Research and Development Strategies for Swedish Companies in the Farm Machinery Industry* (Uppsala: Institute for Economics and Statistics, 1978), and H. Nystrom and B. Edvardsson, *Research and Development Strategies for Four Swedish Farm Machine Companies* (Uppsala: Institute of Economics and Statistics, 1978).
10. R.G. Cooper, "The Impact of New Product Strategies," *Industrial Marketing Management* 12 (1983): 243-56. See also R.G. Cooper, "The Performance Impact of Product Innovation Strategies," *European Journal of Marketing* 18 (1984): 1-54, R.G. Cooper, "How New Product Strategies Impact on Performance," *Journal of Product Innovation Management* 1 (1984): 5-18, R.G. Cooper, "The Strategy-Performance Link in Product Innovation," *R & D Management* 14 (Oct. 1984): 247-59, R.G. Cooper, "New Product Strategies: What Distinguishes the Top Performers," *Journal of Product Innovation Management* 2 (1984): 151-64, R.G. Cooper, "Industrial Firms' New Product Strategies,"

Journal of Business Research 13 (April 1985): 107-21, and R.G. Cooper, "Overall Corporate Strategies for New Product Programs," *Industrial Marketing Management* 14 (1985): 179-83.

11. C.M. Crawford, "New Product Failure Rates — Facts and Fallacies," *Research Management* (Sept. 1979): 9-13.

12. R.C. Bennett and R.G. Cooper, "The Product Life Cycle Trap," *Business Horizons* (Sept.-Oct. 1984): 7-16.

13. T.H. Levitt, "Marketing Myopia," *Harvard Business Review* 38 (July-Aug. 1960: 45-86.

14. I.H. Ansoff, *Corporate Strategy* (New York: McGraw-Hill, 1965), at 103.

15. T.J. Peters and R.H. Waterman Jr., *In Search of Excellence* (New York: Harper & Row, 1982).

16. Corey, "Key Options in Market Selection."

17. D.F. Abell, *Defining the Business* (Englewood Cliffs, N.J.: Prentice-Hall, 1980).

18. Crawford, "Defining the Charter."

19. Cooper, "Strategic Planning."

Appendix A

1. For further information, see R.G. Cooper, "An Empirically Derived New Product Project Selection Model," *IEEE Trans. Engineering Management* EM-28 (Aug. 1981): 54-61; R.G. Cooper, *A Guide to the Evaluation of New Industrial Products for Development* (Montreal: Industrial Innovation Center, 1982); R.G. Cooper, "Selecting Winning New Product Projects: Using the NewProd System," *Journal of Product Innovation Management*, no. 2 (1985): 34-44.

Appendix B

1. D.R. Augood, "Review of R & D Evaluation Methods," *IEEE Trans. Engineering Management* EM-20 (Nov. 1973): 114-20.

2. An excellent review of financial indices is provided ibid., including reviews of measures proposed by Pacifico, Teal, and Disman.

3. R.G. Cooper, *A Guide to the Evaluation of New Industrial Products for Development* (Montreal: Industrial Innovation Center, 1982).

Index

Abell, D.F., 234
Advertising
  decisions, 208-12
  frequency, 211
  message, 211
  objectives, 209
  plan, 208-13
  dealing with the agency, 208-12
  testing, 211
Alternative organizational structures, 42
Ansoff, I.H., 233-34
Arenas, 218-21
  defining, 232-35
  identification, 235-38
  selection criteria, 238-43
Assessing, 240-43
Assessor, pre-test market model, 172
Attrition curve, 16, 17, 18
Attrition rate, 31, 35, 36, 49
Audit, marketing, 190
Audit, new-product practices, 32

Bailey, E.L., 19
Balanced strategy, 223, 225-30
Barriers to idea-generation, 84, 93
Basadur, M., 88, 89
BASES, pretest market model, 172
Benefit measurement models, 101, 103, 104-21
Benefit segmentation, 145-51, 194-95
Benefit-contribution techniques, 105, 108, 109, 253-56
"Better mousetrap", 21
Booz Allen & Hamilton, 6, 7, 11, 12, 13, 16, 221
Boston Consulting Group model, 202
Brainstorming, 71, 86-90
Brand ratings, 138-40
Breakeven analysis, 101
Burke, J.E., 24
Business position, in arenas, 238-43
Buyer behavior, 189

Capital budgeting, 101, 180

Capital expenditures, 12-13
Checklists
  in screening, 41, 104, 105, 109, 110
  advantages, 110
Choice criteria, 142-51
Classes of new products, 5-7
Clipping services, 77
Comparative approaches, 105, 106
Competitive positions, 141-45
Components of risk, 43
Concept, 123-59
  defintion, 57, 133-59
  development, 59, 151-53
  evaluation, 60, 133, 157-58
Concept identification, 58, 133-52
  market study, 133-52
  objectives, 134
  research design, 135-38
Concept test, 59, 133, 153-54
  market study, 153-57
  design, 154-56
  using the results, 156, 157
Conference Board, 8, 10, 16-20, 82, 83, 104
Conservative bets, 240-43
Contribution profit, 205-7
Contribution to profit, 10, 11, 35
Contribution to sales, 8-12
Cooper, R.G., 21, 28, 29, 30, 43, 46, 47, 50-53, 124, 165, 185, 225, 237
Corey, E.R., 234
Cost ratio, 253
Cost reductions, 6, 7
Cost-oriented pricing, 200-1
Crawford, C.M., 15, 80-81, 127, 223, 234
Creativity methods, 71, 86-90, 93
Criteria for market selection, 196
Customer hot line, 73
Customer panels, 72
Customer surveys, 72
Customer tests, 61, 62, 162-71
  after development, 166-71
  during development, 162-63
  need for, 171

Dean, B.V., 70
Defining the Business, 234
Dependence on new products, 9-10
Development, 60
Dimensions of newness, 5
Discounted cash flow, 101, 104, 180, 254

Economic analysis: problems, 103
Economic models, 101, 103, 104
Edvardsson, B., 222
Evaluation, 40, 41, 54, 95-121, 247-56
Evaluative factors, dimensions, 141-43
Expenditures, 32, 40
Extended trials, 169-70

Factors in success, 24-31, 36-42
Failure rate, 15-17, 35, 36
Failure types, 21
Financial analysis, 158, 180, 214-15
Financial indices, 105, 108, 109, 253-56
Financial plan, 214-15
Flow lines, 71
Focal point, 68, 69
Focus groups, 71, 72

Gambling rules, 44-47
Game plan, 35, 48-65
  benefits of, 64-65
Gemini 2000, 3-5, 21, 39, 196
Gemini Inc., 3-5, 19
Generation of new-product ideas, 67-94, 244

High-budget diverse strategy, 225, 226-27
Hopkins, D.S., 10, 17, 19
Hot line, customer, 73

Idea-generating system, 49, 68-94, 220
Idea-generation, 67-94, 244
Idea submissions (unsolicited), 82-83
Ideal new-product process, 38
Ideal product, 133, 134, 143-51
Ideas, new-product, 67-94, 244
Ideation, 87-89
Impact analysis, in screening, 119
Importance of product innovation, 8-10
In Search of Excellence, 233
In-house product testing, 61, 161-62
Incremental commitment, 54
Incrementalized process, 45, 46
Initial screening, 98-121, 247-52
Innovation factors, 12, 13
Intent-to-purchase measurement, 154, 155, 168
Interest measurement, 154, 155
Interfaces
  R & D and marketing, 42

Internal assessment, 191
Isofab Inc., 56

Katz, L.M., 103
Killer phrases, 89
Kreilling, J., 71

Launch, 41, 63
Lessons, 35-42
Levitt, T.H., 233-34
Liking measurement, 154, 155
Little, B., 46, 47
Low-budget conservative strategy, 225, 226, 228-30
Loyalty segmentation, 194

Macroenvironmental analysis, 190
Madique, M.A., 27
Mail surveys, 258-60
  difficulties with, 258-59
Market
  analysis, 189-90
  assessment, 55, 56
  attractiveness, in arenas, 238-43
  attractiveness, in screening, 115
Market factor segmentation, 194
Market orientation, 28, 29, 30, 38, 39, 49
Market pricing, 200-1
Market research, 101-4, 133-57, 204-5, 257-61
  for pricing, 204-5
  models, in screening, 101-4
Market segmentation, 192-95
  bases, 193-95
Market segments, 145-51, 189
Market selection criteria, 196
Market-pull ideas, 49
Marketing audit, 190
Marketing Myopia, 233
Marketing objectives, 185-87
  role of, 185-86
Marketing plan, 41, 42, 60, 183-245
Marketing planning
  process, 42, 183-245
  timing, 183-84
Marquis, D.G., 24, 25
Me-too product, 21, 39
Media plan, 210-11
More, R.A., 43
Moving-target situation, 126
Multidisciplinary inputs, 42
Multi-step new-product process, 36
Multiple evaluation criteria, 100
Must criteria
  in arena selection, 238
  in screening, 54, 90, 110, 112-14, 116-18
Myers, S., 24, 25
Mystery shopping, 75

New corporate practices, 31, 32
New products
    arenas, 210-21, 232-43
    attributes, 57, 127, 198-99
    benefits, 4, 39, 57, 127, 198-99
    concept, 123-59
    development, 58, 60
    failures, 18-23
    features, 127, 198-99
    objectives, 219, 230-32
    performance, 16-17, 223-24
    process, 20, 25, 31, 32, 46, 48-65
    strategy, 17, 31, 217-45
    superiority, 28, 29, 39, 57
    testing (in-house), 61, 161-62
New-product departments, 43
New-product game plan, 35, 48-65
New-to-the-world products, 6
NewProd, 27-31, 247-52
    model, 247-52
    questionnaire, 248-52
No-target situation, 126
Nystrom, H., 222

Observation of customers, 73
Obstacles to product development,
    12-13
Odds of failure, 15
Official Gazette, 77
Ogilvy, D., 210
Organizational structure, 32, 42

"Paralysis by analysis" 37, 38
Patents, 77
Payback period, 101, 104
Penetration pricing, 202
Percentage sales by new products,
    9-12, 17, 35, 224
Performance
    measurement, 32
    objectives, 232
    of new-product programs, 17, 31-32,
    221-30
Personal interviews, 260-61
Peters, T.J., 233
Pilot production, 62
PIMS studies, 202, 221-22
Population defining, 135-37
Portfolio-selection models, 101, 102,
    103
Positioning maps, 141-49
Post-launch evaluations, 63
Pre-commercialization business
    analysis, 63, 180
Pre-development
    activities, 32
    decisions, 123-27
Pretest market, 62, 172, 173
    problems with, 173
    reasons for, 173

Preference measurement, 154, 155
Preference tests, 61, 62, 162, 167-70
    cautions, 167-71
Preliminary assessment, 55, 128-32
    market assessment, 56, 128-31
    project evaluation, 128, 131-32
    technical assessment, 128, 131
Price crunch, 22
Price sensitivity, 168
Pricing
    decisions, 200-8
    effect of corporate strategy on, 203
Process model, 48-65
Product additions, 6, 7
Product advantage, in screening, 115
Product champion, 25, 27, 37
Product improvements, 6, 7
Product innovation charter, 217-45
    definition, 218-19
    developing, 230-43
    implementing, 243-45
    importance, 221-22
    objectives, 219, 230-32
    role, 219, 221, 222
Product life-cycle trap, 93, 230
Product positioning, 127, 141-49,
    197-99, 201, 209-10
Product specifications, 57, 127, 198-99
Product strategy, 57, 197-99
Product superiority, 28, 29, 39, 57
Product usage, 194
Product value, 39, 203-4
Profile charts, models, 105-7
Project evaluation, 40, 41, 54, 95-121,
    247-56
    difficulties, 95-98
Project feasibility, 55
Promotional pricing, 207-8
Protocol, 57, 60, 127, 128, 158, 161,
    197-99

Q-sort, 105, 106
Questionnaire
    attribute importance, 137
    concept test, 155
    customer perceptions, 136
    ideal product, 137
    NewProd, 248-52

R & D spending, 8, 9, 12, 219
    by country, 8
    by industry, 8, 9
    percentage of sales, 8, 12, 35
Reach, in advertising, 211
Reasons for performance, 17-33,
    221-30
Reducing risk, 43-47
Relative expenditure, 12-13
Relative preferences, 138
Repositionings, 6, 7

Resource allocation decisions, 96, 98, 101, 102
Resource allocation, spending, 12, 13, 16, 17
Resource deficiencies, 22
Resource planning, 220-21, 244-45
Response bias, 260
Response rate, 259-60
Return on investment, 101, 104
Risk, 43-47
Risk management, 43-47, 49
Roberts, R.W., 24
Role objectives, 231-32
Rothwell, R., 37
Rubenstein, A.H., 26

Sales force decisions, 212-13
Sample selection, 135-37
SAPPHO, Project, 25, 26, 37
Scoring models, 41, 104, 105, 110-21, 247-52
    advantages, 112
    problems with, 112-13
Screening, 41, 54, 98-121, 220, 244, 247-56
    errors, 99
    model, 247-52
    briefing session, 118-19
    methods, 102-4
    requirements, 99-100
    using a system, 117-20
Selling task, 213
Sensitivity analysis, 180
Should criteria, in screening, 55, 90, 11, 115-18, 132
Situation size-up, 187-91
Skimming, pricing, 202
Souder, W.E., 104, 105
Sources of ideas, 70-94
    competitors, 74, 75
    customers, 70-74
    inside the company, 84-93
    brokers, 78
    licensing shows, 78
    opportunities form, 92
    private inventors, 79-81
    suggestion schemes, 85-86
    sales force, 91, 92
    universities, 81
    suppliers, 79
    trade shows, 75
Stanford Innovation Project, 27
Statistical Abstract of the U.S., 9
Stepwise model, 36, 48-65
Strategy, 217-218
    technologically deficient, 225-26
    technologically driven, 224-26, 228
Strategy types, 222-30
Success factors, 5, 24, 25, 36-42
Success rate, 16, 17, 18, 224

Successful cases, 25
Superiority of product, 28, 29, 39, 57
Supporting elements, 200
Survey methods, 257-61
Synergy, in screening, 115
    in strategy, 233

Target market, 41, 126, 201, 209, 210
    defining, 57, 192-96
    selecting, 195-96
Technical assessment, 55, 56
Technological fit, 29, 30
Technological synergy, 29, 30
Telephone surveys, 257-58
TEMP (pre-test market model), 172
Test market, 62, 172, 174-80
    locations, 175
    designing, 175-76
    industrial products, 175
    measuring results, 176
    problems with, 177-80
    reasons for, 174
    when to test, 177-80
Testing, 61, 161-81
    advertising, 211
    market acceptance, 172-80
    need for, 161
Timing, 20
Top performers, 17, 227-28
Tracking projects, 120
Trade publications, 76
Trade shows, 75
Trend analysis, 190
Trial, 62, 171-81
Trial production, 62
Trial sell, 62, 175
Types of failures, 21

U.S. patent office, 77
Uncertainties, 43-47
Uncertainty of information, 99
Unique product, 28, 29, 39, 57
Unsolicited ideas, 79-81
Up-front activities, 32, 39, 40
User-developers, 73, 74
User-innovators, 74
Users' panel, 163

Value of a product, 39, 203-4
Value-in-use, 204
Variable costs, 205-7
Venture teams, 43
Volume segmentation, 194
Von Hippel, E.A., 70, 71, 73, 74

Waterman, R.H., 233

Zirger, B.J., 27